美国数学会经典影印系列

出版者的话

近年来，我国的科学技术取得了长足进步，特别是在数学等自然科学基础领域不断涌现出一流的研究成果。与此同时，国内的科研队伍与国外的交流合作也越来越密切，越来越多的科研工作者可以熟练地阅读英文文献，并在国际顶级期刊发表英文学术文章，在国外出版社出版英文学术著作。

然而，在国内阅读海外原版英文图书仍不是非常便捷。一方面，这些原版图书主要集中在科技、教育比较发达的大中城市的大型综合图书馆以及科研院所的资料室中，普通读者借阅不甚容易；另一方面，原版书价格昂贵，动辄上百美元，购买也很不方便。这极大地限制了科技工作者对于国外先进科学技术知识的获取，间接阻碍了我国科技的发展。

高等教育出版社本着植根教育、弘扬学术的宗旨服务我国广大科技和教育工作者，同美国数学会（American Mathematical Society）合作，在征求海内外众多专家学者意见的基础上，精选该学会近年出版的数百种专业著作，组织出版了“美国数学会经典影印系列”丛书。美国数学会创建于 1888 年，是国际上极具影响力的专业学术组织，目前拥有近 30000 会员和 580 余个机构成员，出版图书 3500 多种，冯·诺依曼、莱夫谢茨、陶哲轩等世界级数学大家都是其作者。本影印系列涵盖了代数、几何、分析、方程、拓扑、概率、动力系统等所有主要数学分支以及新近发展的数学主题。

我们希望这套书的出版，能够对国内的科研工作者、教育工作者以及青年学生起到重要的学术引领作用，也希望今后能有更多的海外优秀英文著作被介绍到中国。

高 等 教 育 出 版 社

2016 年 12 月

AMS
AMERICAN MATHEMATICAL SOCIETY
美国数学会经典影印系列

Lectures and Problems: A Gift to Young Mathematicians

讲义与问题：给青年数学家的礼物

V. I. Arnold
Translated by Dmitry Fuchs and Mark Saul

中国教育出版传媒集团
高等教育出版社 · 北京

Contents

Preface to the English Edition

Vladimir Arnold was one of the great mathematical minds of the late 20th century. He did significant work in many areas of the field. On another level, Russian mathematicians have a strong tradition of writing for, and even directly teaching, younger students interested in mathematics. The present volume contains some examples of Arnold's contributions to the genre.

"Continued Fractions" takes a common enrichment topic in high school math and pulls it in directions that only a master of mathematics could envision. While it exemplifies for the student the kind of generalization and abstraction that mathematicians routinely engage in, it does so in a completely non-routine way. The essay also has a powerful lesson for all of us. The author claims to have set out to invent a completely useless (i.e., inapplicable) mathematical construct, yet found that people came to his door asking about it because it was just what they needed for a particular application. Mathematicians, it seems, do more than build a better mousetrap. They seem to invent new creatures to be trapped.

In "Geometry of Complex Numbers, Quaternions, and Spins" the context is physics, yet Arnold artfully extracts the mathematical aspects of the discussion in a way that students can understand long before they master the field of quantum mechanics.

"Euler Groups and Arithmetic of Geometric Progressions" treats a similar enrichment topic, but it is rarely treated with the depth and imagination lavished on it here. Arnold sets it in a mathematical context, bringing to bear numerous tools of the trade and expanding the topic way beyond its usual treatment.

"Problems for Children 5 to 15 Years Old" must be read as a collection of the author's favorite intellectual morsels. Many are not original, but all are worth thinking about, and each requires the solver to think out of his or her box. Dmitry Fuchs, a long-term friend and collaborator of Arnold, provided his solutions to some of the problems. Readers are of course invited to select their own favorites, and construct their own favorite solutions.

In reading these essays, one has the sensation of walking along—sometimes being dragged along—a simple footpath that is found to ascend a mountain peak, and being shown a vista whose existence one could never suspect from the ground.

Arnold's style of exposition is unforgiving. The reader—even the professional mathematician—will find paragraphs that require hours of thought to unscramble. In some cases, Arnold collapses an argument into a few sentences that might take up several pages in another style of exposition. In other cases, he gives an intuitive argument in place of a rigorous one, leaving the reader to construct the latter. He probably felt that the real work was done on the intuitive level, and that his teaching would be the more effective if he left the tidying up to the student. The reader must have patience with the ellipses of thought and the leaps of reason. They are all part of Arnold's intent.

These notes were often gathered from the field, and we have corrected numerous misprints and small errors in notation. We have also given several extensions—in Arnold's own style—to the work in "Editors' Comments". At the same time, we have striven to deliver intact the style of the essays. Arnold's mind leaps from peak to peak, connecting disparate areas of mathematics, all (or most) accessible to the student with an advanced high school education. And yet there is a unity to each essay, a flow from very simple questions to deep intellectual inquiry, and sometimes right to the edge of our knowledge of mathematics.

We hope that we have preserved this coherence, but also the excitement of the work, the sharp, jagged edges and breathtaking jumps that characterize the author's thinking.

It is our pleasure to acknowledge the contributions of several colleagues to this work. In particular, Sergei Gelfand, at the American Mathematical Society, kept us on track at several key junctures. Paul Zeitz made important contributions to the work of translation. James Fennell sedulously proofread the manuscript and corrected the TeX files. We would also like to thank the students of the Gradus ad Parnassum math circles at the Courant Institute, who gave us feedback about several sections. Much of this work was supported by a generous grant from the Alfred P. Sloan Foundation.

Mark Saul
June 2015

Part 1

Continued Fractions

Continued Fractions

What is a Continued Fraction?

The theory of continued fractions is one of the oldest in mathematics. To explain what a continued fraction is, we start with a simple example. Let us take the fraction $\frac{10}{7}$. The largest integer not greater than this fraction is 1:

$$\frac{10}{7} = 1 + \frac{3}{7} \quad \left(\frac{3}{7} < 1\right).$$

Let us "turn $\frac{3}{7}$ upside down":

$$\frac{10}{7} = 1 + \frac{3}{7} = 1 + \frac{1}{\frac{7}{3}}.$$

The largest integer not exceeding $\frac{7}{3}$ is 2. So we have:

$$\frac{10}{7} = 1 + \frac{3}{7} = 1 + \frac{1}{\frac{7}{3}} = 1 + \cfrac{1}{2 + \cfrac{1}{3}}.$$

This is the *continued fraction* for the number $\frac{10}{7}$, which, among other things, provides very good approximations. The fraction $\frac{10}{7}$ is rather close to 1, but if you want more precision, it is closer to $1+\frac{1}{2}$, and the expression $1+\cfrac{1}{2+\cfrac{1}{3}}$ gives the exact value.

We can represent any number in this way. If the number is irrational, then this process will continue indefinitely without terminating. For a rational number, the continued fraction representation will be finite.

The continued fraction for π

What is π?

In the *Proceedings of the USSR Academy of Sciences* for 1935 I read two papers by biologists, both of whom mentioned the number π. One article was entitled "On the Pecking of Woodpeckers", and the other was called, "On the Spouting of Whales". This last article mentioned the following problem in whale hunting. Suppose you have noticed the spout of a whale from a distance. You want to know whether it is worth the effort to go after this whale, or if the quantity of meat you would obtain is not significant. For this, we must understand how the spout of the whale depends on the volume of the animal's body. Therefore the article includes a formula for the volume of a whale: $V = \pi r^2 \ell$, where r is half the width of the whale's body and ℓ is its length (the whale is assumed to be cylindrical). The only difficulty in explaining this formula to whalers is the number π, which the article defines as "*... a constant, which for Greenland whales is equal to 3.*" But for other species, clearly, one must use a different value.

Approximations to π were known to the ancients. Here, for example, is a very good approximation, attributed to Archimedes, but which was known even before his time: $\pi \approx \frac{22}{7} = 3\frac{1}{7}$. In fact, this is actually the beginning of the continued fraction for π. This continued fraction is infinite, and if we go further and further out, we can get better and better approximations (see Page 5).

Note that the numerator of the fraction $\frac{22}{7}$ is a two-digit number while the denominator has one digit, and that the accuracy of the approximation is to three decimal places (see Table on Page 5, (a)). We can get six decimal places of accuracy by truncating the continued fraction further down (c). This new approximation is the ratio of two three-digit numbers. Here is a rule that can help us remember this fraction: just write down 113355, break it into two three-digit numbers, and divide the larger by the smaller. We get:

$$\pi = 3 + \cfrac{1}{7 + \cfrac{1}{16}} = \frac{355}{113}.$$

In my view, mathematics and physics are parts of the same experimental science. When the experiments cost billions of dollars we call this science physics. When they are cheap, we call it mathematics. Furthermore, mathematics is a unified whole that must not be divided into algebra, geometry, etc. In particular, the sort of computations that we have been doing arose in the construction of the calendar, when the ratio of the solar year and the period of the moon was expressed as a fraction. The closest approximation to this ratio is 12 (like 3 for π). Various corrections were introduced: first, leap years. Then the Gregorian calendar corrected the Julian, not just with leap years, but with another correction every 100 years, and another every 400 years, and another....

These commensurability adjustments turned out to be particularly important as celestial mechanics and astronomy developed. For example, the

$$\pi = 3 + \cfrac{1}{\underset{(a)}{7} + \cfrac{1}{\underset{(b)}{15} + \cfrac{1}{\underset{(c)}{1} + \cfrac{1}{\underset{(d)}{292} + \cfrac{1}{\underset{(e)}{1} + \cfrac{1}{\underset{(f)}{1} + \cfrac{1}{\underset{(g)}{1} + \cfrac{1}{\underset{(h)}{2} + \cfrac{1}{\underset{(i)}{1} + \cfrac{1}{\underset{(j)}{3} + \cfrac{1}{\underset{(k)}{1} + \cfrac{1}{\underset{(l)}{14} + \cfrac{1}{\underset{(m)}{2} + \cfrac{1}{\underset{(n)}{1} + \ddots}}}}}}}}}}}}}}$$

	Rational approximation of the number $\pi \approx 3.14159265358979324$	The number of decimal places matching
(a)	$\frac{22}{7} \approx 3.\underline{14}2$	2
(b)	$\frac{333}{106} \approx 3.\underline{1415}1$	4
(c)	$\frac{355}{113} \approx 3.\underline{141592}9$	6
(d)	$\frac{103\,993}{33\,102} \approx 3.\underline{141592653}0$	9
(e)	$\frac{104\,348}{33\,215} \approx 3.\underline{141592653}9$	9
(f)	$\frac{208\,341}{33\,215} \approx 3.\underline{141592653}467$	9
(g)	$\frac{312\,689}{99\,532} \approx 3.\underline{141592653}6$	9
(h)	$\frac{833\,719}{265\,381} \approx 3.\underline{14159265358}1$	11
(i)	$\frac{1\,146\,408}{364\,913} \approx 3.\underline{1415926535}91$	10
(j)	$\frac{4\,272\,943}{1\,360\,120} \approx 3.\underline{141592653589}4$	12
(k)	$\frac{5\,419\,351}{1\,725\,033} \approx 3.\underline{141592653589}8$	12
(l)	$\frac{80\,143\,857}{25\,510\,582} \approx 3.\underline{14159265358979}26$	14
(m)	$\frac{165\,707\,065}{52\,746\,197} \approx 3.\underline{141592653589793}4$	15
(n)	$\frac{245\,850\,922}{78\,256\,779} \approx 3.\underline{141592653589793}16$	15

commensurability of the periods of Jupiter and Saturn about the sun (approximately 2 : 5) leads to a very strong perturbation, which knocks the planets out of their orbits. This is the so-called "great inequality" in the motions of Jupiter and Saturn, which has a period of about 800 years. In computing such periods, continued fractions and their associated approximations have great value and required serious developments of mathematical apparatus. These developments rather quickly led to the understanding that this arithmetic was in fact geometry.[1]

Below I present several vignettes from the theory of continued fractions, and show the geometric meaning behind them, based on empirical methods.

This sort of geometry became popular about 100 years ago, thanks to the work of the great mathematician Hermann Minkowski, who called it the "Geometry of Numbers". Minkowski's predecessors used this theory without naming it, and so are forgotten.

The Geometric Theory of Continued Fractions

At the foundation of Minkowski's geometry of numbers lies a simple piece of graph paper: a plane on which a coordinate grid has been drawn. Let us look at the line $y = \alpha x$; for example, let $\alpha = \frac{10}{7}$. If α is a rational number, then there will be integer points on this line other than just the origin. In our case, the line will go through the point $(7, 10)$.

It turns out that the construction of a continued fraction for the number α is connected with the problem of finding integer points which lie close to our line.

In particular, there is a geometric algorithm, which was explained to me by the great Russian mathematician Boris Nikolaevich Delaunay when I was a freshman in college. He gave this algorithm the expressive name "stretching noses". This algorithm allows us to construct the integer points closest to a line one after another and at the same time obtain a continued fraction.

The nose stretching algorithm

Let $\overrightarrow{e_1} = (1, 0)$ and $\overrightarrow{e_2} = (0, 1)$ be the standard unit basis vectors, between which our line is located (Figure 1). We start adding $\overrightarrow{e_2}$ to the vector $\overrightarrow{e_1}$ as long as the sum doesn't cross our line. In other words, we need to find the largest non-negative integer number a_0 such that the endpoint of the vector $\overrightarrow{e_3} = \overrightarrow{e_1} + a_0\overrightarrow{e_2}$ is still below our line. In the case shown, $a_0 = 1$.

Let us proceed. To construct the vector $\overrightarrow{e_4}$, we add to the vector $\overrightarrow{e_2}$ the vector $\overrightarrow{e_3}$ (which we have already constructed) multiplied by the coefficient a_1. We choose a_1 in such a way that we don't cross the line; that is, the vector $\overrightarrow{e_4}$ stays above the line, but if we add $\overrightarrow{e_3}$ to it one more time, it would cross the line. As we see, here $a_1 = 2$.

The vectors get longer and longer, hence the name, "stretching noses".

[1] All of this ancient body of knowledge (including the "Euclidean Algorithm", the theory of "Pythagorean triples" such as $3^2 + 4^2 = 5^2$, and a rigorous theory of irrational numbers) was known to ancient Egyptian astronomers thousands of years before Pythagoras, Euclid, or Eudoxus, who disseminated these ideas to the ancient Greeks.

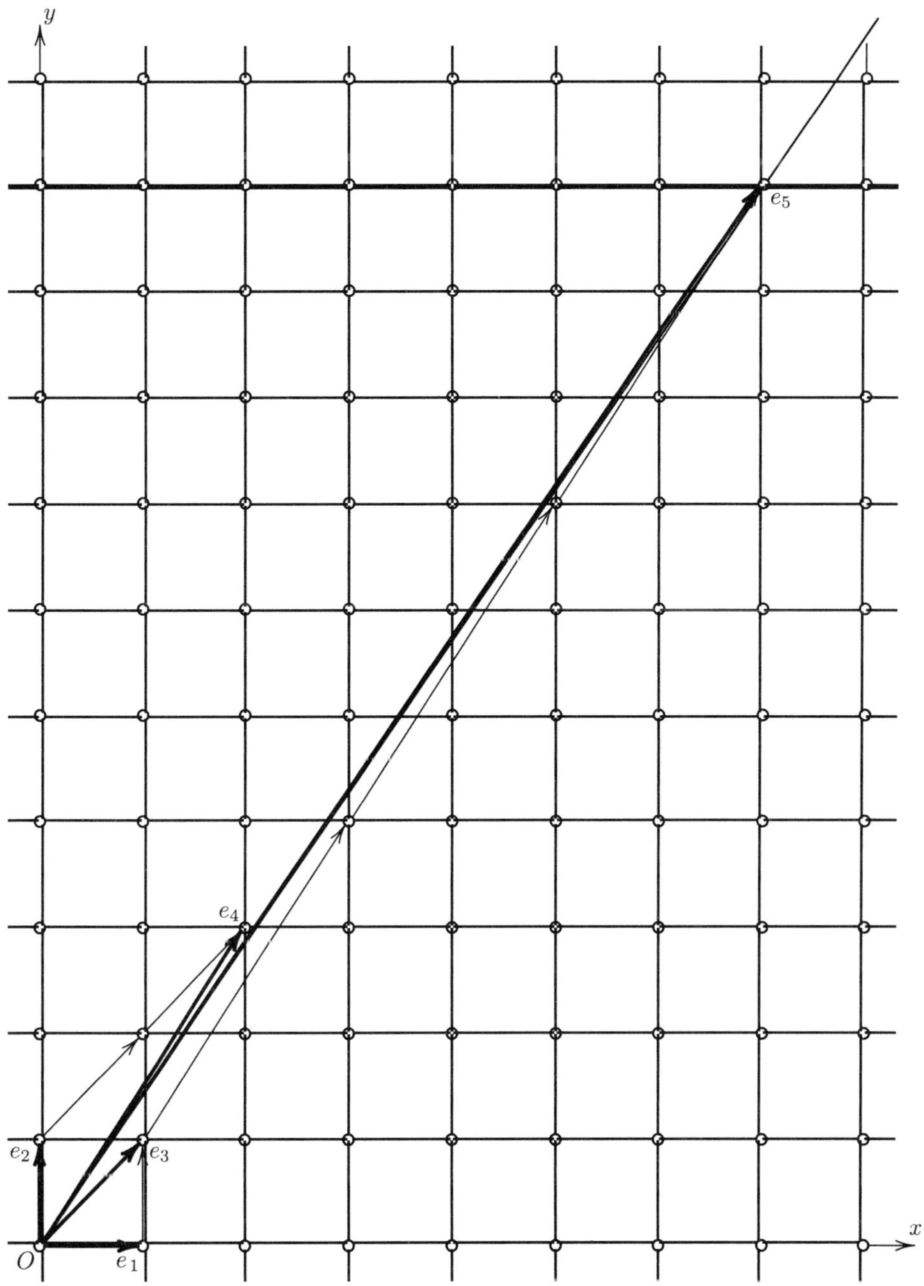

FIGURE 1. Nose stretching.

Continuing, $\overrightarrow{e_5} = \overrightarrow{e_3} + a_2\overrightarrow{e_4}$. When we take $a_2 = 3$, we land directly on the line. Hence $a_0 = 1$, $a_1 = 2$, $a_2 = 3$, and

$$a_0 + \cfrac{1}{a_1 + \cfrac{1}{a_2}} = 1 + \cfrac{1}{2 + \cfrac{1}{3}} = \frac{10}{7}.$$

We can prove that this algorithm always gives the same integers $a_0, a_1, a_2, \ldots$ that we obtain in representing α with a continued fraction. The points we obtain immediately give us the terms of the continued fraction.

The proof is not complicated. The main thing is that a line with the equation $y = Ax$ with respect to some coordinate system will have the equation $x = \frac{1}{A}y$ with respect to the coordinate system with abscissa and ordinate interchanged. Also, a line with the equation $y = Ax$ with respect to the coordinate system with the basis vectors $\vec{e}$ (on the x axis) and $\vec{f}$ (on the y axis) is also given by the equation $z = Bw$, where $A = a + B$, with respect to the coordinate system with the basis vectors $\vec{e} + a\vec{f}$ (on the w axis) and $\vec{f}$ (on the z axis). The continued fraction is obtained by a successive application of these two (obvious) facts. (See [**EC1**].)

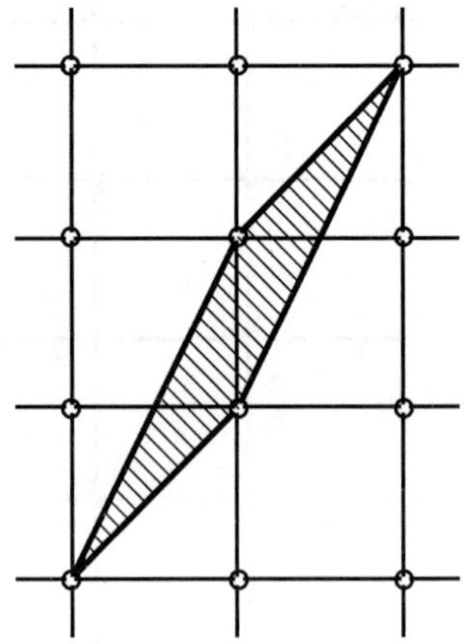

FIGURE 2.

Two lemmas on the geometry of numbers

I will now prove two fundamental lemmas, which form the foundation of the geometry of numbers.

Lemma. *On a coordinate grid we consider an "empty" parallelogram whose vertices are lattice points; i.e., the parallelogram has no other lattice points either inside it or on its sides (for example, as in Figure 2). Then the area of this parallelogram is 1.*

Of course, it is not hard to compute the area of a parallelogram. But I will show how a physicist might prove this lemma. From a mathematician's point of view, this is not a proof: it does not use axioms.

In his *Confessions*, Jean-Jacques Rousseau recounts how, when he started school, and learned how to simplify parentheses, he derived a wonderful formula–the formula for the square of a sum: $(a + b)^2 = a^2 + 2ab + b^2$. But, although he discovered this himself, and didn't doubt that he had removed the parentheses correctly, he couldn't really believe this formula–until he found another proof, without parentheses. Here is that proof: we divide a square with side $a + b$ into four parts (Figure 3), from which it is clear that its area is equal to $a^2 + ab + ba + b^2$. After this, all his doubts disappeared.

I call such a proof a "physics proof", and in my view, these are the only real, convincing proofs, and the only ones which render mathematics comprehensible. No removal of parentheses, no algebra is really convincing. There might be errors in the algebra, and even computer programs can fail.

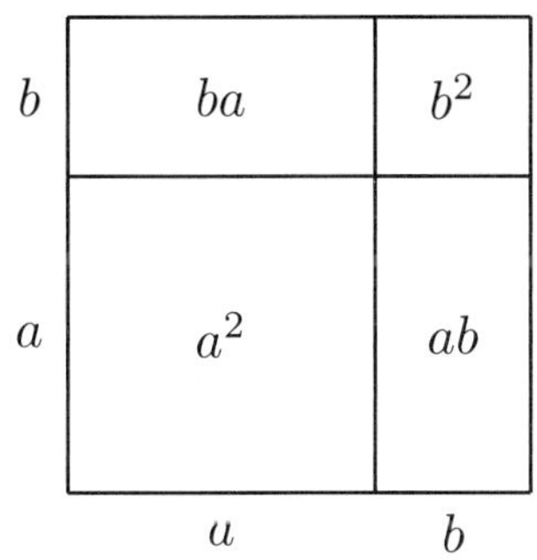

FIGURE 3.

So I will prove this lemma with "physics", *à la* Rousseau.

Proof of Lemma. Translating our parallelogram by all possible combinations of its spanning vectors, we can cover the entire plane with congruent parallelograms, in the same manner that the plane is covered by unit squares formed by the coordinate grid lines (Figure 4).

Consider a set in the plane with a large area A and first count how many parallelograms it contains, then how many lattice points it contains. Let the area of the parallelogram be S. Then if A is very large, the number of parallelograms is approximately equal to A/S. (This region may not consist of whole parallelograms, so the value is not exact. However, we could take a set consisting of whole parallelograms and then the value would be exact). It is clear that the number of lattice points is approximately A.

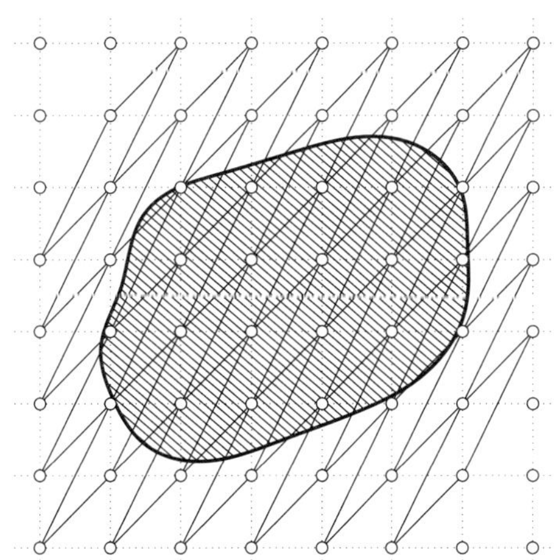

FIGURE 4.

Now we will count the number of lattice points contained in our region in a different way. Each parallelogram contributes 4 lattice points (its vertices), but now we are counting each vertex four times, and if we count all the vertices of each parallelogram, we will get a result that is four times bigger than the number of lattice points in all. Thus, the number of lattice points and the number of parallelograms are equal. Thus $A \approx A/S$ for a very large A. This means that $S = 1$.

Remark. This argument can be easily generalized to the case where the parallelogram with lattice points as vertices also contains k internal lattice points and l lattice points on its boundary. The area of such a parallelogram is $S = 1 + \beta k + \gamma l$. The reader is invited to find the coefficients β and γ on his own and thus get the answer (which can be empirically verified using small values of k and l).

Lemma (Area formula for parallelograms). *Consider a parallelogram that is spanned by vectors with coordinates* (a, b) *and* (c, d) *(the numbers* a, b, c, d *are not necessarily integers). We will consider the area of the parallelogram to have a positive sign if the direction of rotation from the first to the second vector is the same as the direction of rotation from the* x*- to the* y*-axis, and a negative sign in the opposite case. Then* $S = \begin{vmatrix} a & b \\ c & d \end{vmatrix}$.

The number $\begin{vmatrix} a & b \\ c & d \end{vmatrix} = ad - bc$ is called the determinant of the matrix $\begin{pmatrix} a & b \\ c & d \end{pmatrix}$.

Proof of Lemma. The area of a parallelogram is a linear function of a vector: if we replace the first vector with a sum of two others, the area of the two new parallelograms formed will add up to the area of the original parallelogram (see **[EC2]**). Furthermore, if the two vectors are interchanged, then the area will change sign (by the definition we have made of signed areas). From these two facts and because the area of a unit square is one, it follows immediately that $S = \begin{vmatrix} a & b \\ c & d \end{vmatrix}$ is the only possible formula.

This is the unique function which is linear in the first argument, linear in the second, antisymmetric (the sign changes when the arguments are interchanged) and equal to one for the two basis vectors (see **[EC3]**).

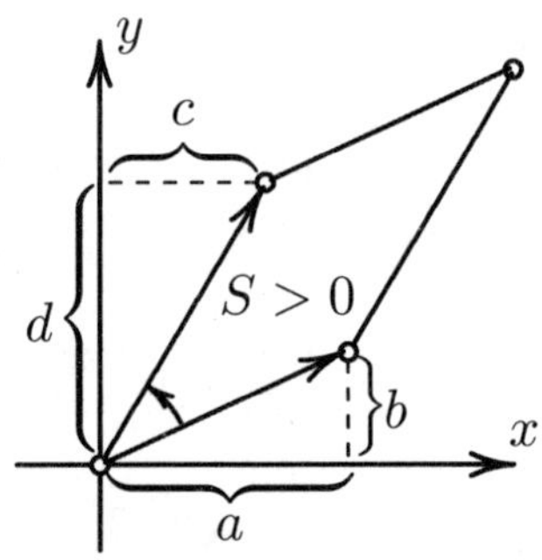

FIGURE 5.

In algebra, this subject area is called the theory of determinants. To enhance the authority of their area, algebraists hide the fact that determinants are simply areas, volumes, etc., by defining them as ghastly polynomials constructed with complex rules, making the theory of determinants completely incomprehensible. But if one starts out by defining the determinant to be area or volume, then all the theorems in the theory of determinants become perfectly obvious, and the proofs, which I call physics proofs, or proofs *à la* Rousseau, can be seen immediately.

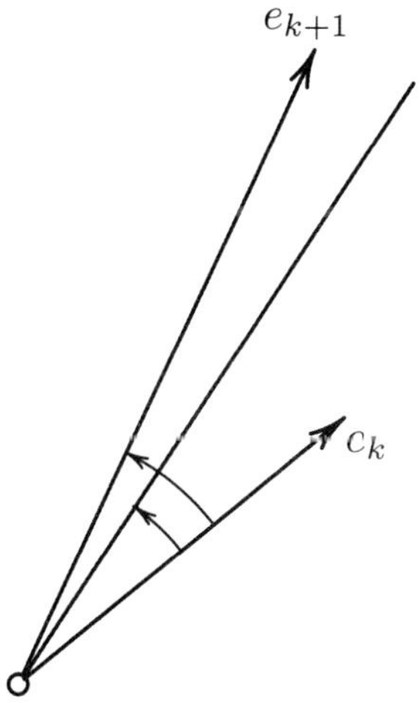

FIGURE 6.

Let us return to our algorithm. The vectors $\overrightarrow{e_1}$ and $\overrightarrow{e_2}$ determine a unit square, so the corresponding determinant is equal to 1. Consider the vector $\overrightarrow{e_3}$. The rotation from $\overrightarrow{e_2}$ to $\overrightarrow{e_3}$ is in the negative direction, and the sides and interior of the parallelogram contain no lattice points. Hence this determinant is equal to -1. Continuing further, we see that the construction at each step is the following. Given a parallelogram (spanned by $\overrightarrow{e_{k-1}}$ and $\overrightarrow{e_k}$), we repeatedly add the vector $\overrightarrow{e_k}$ to its other side ($\overrightarrow{e_{k-1}}$), replacing the first side in the sum ($\overrightarrow{e_{k+1}} = \overrightarrow{e_{k-1}} + a_{k-2}\overrightarrow{e_k}$) and interchanging the sides. The absolute value of the area does not change. Only the sign changes. Let (q_k, p_k) be the coordinates of the vector $\overrightarrow{e_k}$ (q_k and p_k are integers). The area S_k of the parallelogram spanned by vectors $\overrightarrow{e_k}$ and $\overrightarrow{e_{k+1}}$ is equal to

$$\begin{vmatrix} q_k & p_k \\ q_{k+1} & p_{k+1} \end{vmatrix}.$$

Here is the fundamental result of the theory of continued fractions.

Theorem. $S_k = (-1)^{k+1}$ *(for $k \geq 1$).*

Proof. Indeed, we showed that $S_k = \pm 1$ and that the sign changes each time. Thus $S_k = (-1)^k$ or $S_k = (-1)^{k+1}$ for all k. Since, $S_1 = 1$, this proves our theorem.

Corollary. *The fraction $\dfrac{p_k}{q_k}$ is an extremely good approximation for our number α. The formula*

$$\alpha \approx \frac{p_k}{q_k}, \quad k \geq 3,$$

has an accuracy of order $1/q_k^2$.

Proof. We will prove a more precise inequality, from which the result follows. The line $y = \alpha x$ passes through the interior of the parallelogram generated by the vectors $\overrightarrow{e_k}$ and $\overrightarrow{e_{k+1}}$. One of them is below this line, and the other is above it. (Precisely which one depends on the parity of k).

Consequently,

$$\left|\alpha - \frac{p_k}{q_k}\right| \leq \left|\frac{p_{k+1}}{q_{k+1}} - \frac{p_k}{q_k}\right|,$$

because the angle between the line and the vector $\overrightarrow{e_k}$ is no more than the angle between the vectors $\overrightarrow{e_{k+1}}$ and $\overrightarrow{e_k}$ (cf. Figure 5).

Furthermore,

$$\left|\frac{p_{k+1}}{q_{k+1}} - \frac{p_k}{q_k}\right| = \frac{|p_k q_{k+1} - q_k p_{k+1}|}{|q_k q_{k+1}|} = \frac{1}{q_k q_{k+1}},$$

since $|p_k q_{k+1} - q_k p_{k+1}| = |S_k| = 1$ by the theorem proven above, and because q_k and q_{k+1} are positive. Thus

$$\left|\alpha - \frac{p_k}{q_k}\right| \leq \frac{1}{q_k q_{k+1}} < \frac{1}{q_k^2},$$

since $q_{k+1} > q_k$. The precision of the approximation $\alpha \approx \frac{p_k}{q_k}$ is better than $\frac{1}{q_k q_{k+1}}$, and certainly better than $\frac{1}{q_k^2}$. This is why continued fractions give such accurate approximations (see **[EC4]**).

Kuzmin's Theorem

In physics, continued fractions first arose in astronomical investigations. They were used not only to construct calendars but to calculate eclipses, planetary motion, and other periodic phenomena arising in celestial mechanics. In describing the commensurability of different frequencies of periodic motion, such as the Keplerian motion of the planets, astronomers were compelled to find good rational approximations to these numbers, which are generally irrational. It was especially important to find good rational approximations with denominators that were not very large. An approximation that is too close is called a resonance and can lead to strong perturbations of one planet's motion by the others.

Consider the following model. Suppose two planets revolve around a "Sun" along concentric circles in the same direction. If the ratio of the periods of their revolutions around the "Sun" is very close to a rational number, say $\frac{10}{7}$, then these two planets will be close to each other (at the smallest possible distance) near three fixed points (cf. Figure 7). At small distances, as is well known, the pull of gravity is greatest, so the orbits of the two planets will experience strong deformations in only three directions, as if they were pushing each other off their orbit. The situation is completely different if the ratio can be approximated well by a rational number with a large denominator. Suppose, say, this value is $\frac{151}{700}$. Then there will be 549 "points of greatest gravitation", and the mutual interactions between the planets will be "smeared" along the orbit.

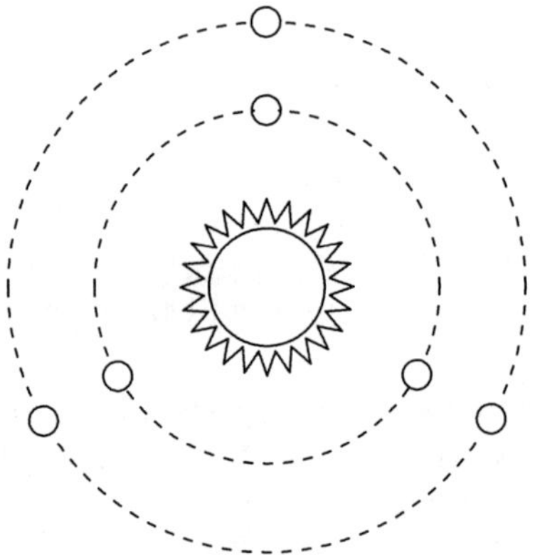

FIGURE 7.

Thus astronomers early on (this was already of interest to Newton and Kepler) asked themselves: What are the values is of these "incomplete quotients" (elements) of a continued fraction? That is, if

$$\alpha = a_0 + \cfrac{1}{a_1 + \cfrac{1}{a_2 + \ddots}},$$

then how large are the terms $a_0, a_1, a_2, \ldots$, if α is just some random real number? If some term, for example, a_2, is very large—say, a million—then the approximation $\alpha \approx a_0 + \dfrac{1}{a_1}$ (which we obtain if we truncate the fraction before a_2) will be incredibly accurate. But if a_2, say, is just 2, then the error will be rather large. Therefore the question of whether the terms grow, and how fast they grow, has real astronomical significance for the fate of the Universe, the fate of the Solar System, and the fate of our civilization.

The first mathematical investigation of this important question probably was done by the astronomer H. Gylden, who published it in the Proceedings (*Comptes Rendus*) of the Paris Academy of Sciences in 1888 [**24**]. I think that this was an experimental work, because astronomers had investigated the ratios of the periods of different planets, large and small, and knew the coefficients a_i of these ratios–not very many, but they knew them. And Gylden produced tables from which it was possible to determine how large the a_i were.

The theorem which answered this question is known as Kuzmin's theorem, although it was apparently proven by the great Swedish mathematician A. Wiman, who published it in 1900 in the Memoirs of the Royal Academy of Sciences of Stockholm [**35**] (R. O. Kuzmin proved it only in 1928). Unfortunately, neither Kuzmin nor anyone else read Wiman's work, because it was 300 pages long. For me this work remains an enigma. I don't know what it contains. I don't know whether it contains a formulation of Kuzmin's theorem or its proof.

A proof of Kuzmin's theorem can be found in A. Ya. Khinchin's book on continued fractions [**27**], which, in fact, is basically devoted to the proof of this theorem.

The fundamental discovery needed for the proof of this theorem was made by Gauss. Although he apparently neither proved nor formulated this theorem, he found an answer: he determined the probability that a term a_i will be equal 1 or 2 or 3, etc. Gauss gave a formula for these probabilities. But how Gauss found this formula, and what he made of it, remains a mystery.

This probability is defined as follows. We take the terms $a_0, a_1, \ldots, a_n$ (all positive integers), and see how many of them are equal to, say, one. Then we divide by n and let n go to infinity. It turns out that this limit exists for almost all α and is the same for all values of α. This number is called the probability p_1 of the appearance of a value of 1.

Kuzmin's theorem asserts that the probability of the appearance of a value of k is given by the following formula:

$$p_k = \frac{1}{\ln 2} \ln\left(1 + \frac{1}{k(k+2)}\right). \tag{1}$$

(The value $\dfrac{1}{\ln 2}$ is a normalizing coefficient that is independent of k; it is necessary in order for the sum of all the probabilities to equal 1.)

If k is a large number, then

$$\frac{1}{k(k+2)} \approx \frac{1}{k^2},$$

which is a small number, and the natural logarithm of the sum of 1 plus a small number is approximately equal to this small number. Consequently, as k grows, p_k decreases as $\dfrac{1}{k^2 \ln 2}$; that is, inversely proportional to the square of k. So when k is large, the probability is small. The greatest probability is for $k = 1$: we have

$$\frac{1}{k(k+2)} = \frac{1}{3}, \qquad \text{so } p_1 = \frac{1}{\ln 2} \ln \frac{4}{3} \approx 0.41.$$

We see that one appears often, almost half the time (cf. the table on page 5).

The Golden Section

An interesting number, known since ancient times, is given by the continued fraction for which all coefficients a_i equal 1:

$$1 + \cfrac{1}{1 + \cfrac{1}{1 + \ddots}}.$$

Denote this number by x. It satisfies the equation

$$1 + \frac{1}{x} = x, \quad \text{or} \quad x^2 - x + 1 = 0,$$

which yields $x = \dfrac{1 \pm \sqrt{5}}{2}$. Since x must be positive, $x = \dfrac{1 + \sqrt{5}}{2} \approx 1.6$.

This number has its own name: the *golden section*. It is a very beautiful number. For example, postcards are made in the form of a rectangle with the ratio of its sides equal to this number. If, from such a rectangle, we cut a square whose sides are equal to the smaller dimension of the rectangle, then the remaining rectangle will be similar to the original one (cf. Figure 8 (a)). This is in fact the condition that the ratio of the sides of the rectangle be the golden section. If we cut off a square again, we will again obtain a rectangle similar to the original (cf. Figure 8 (b)), etc.

I would like to say a few more words about Kuzmin's theorem so it can be considered as an exercise, albeit not an easy one.

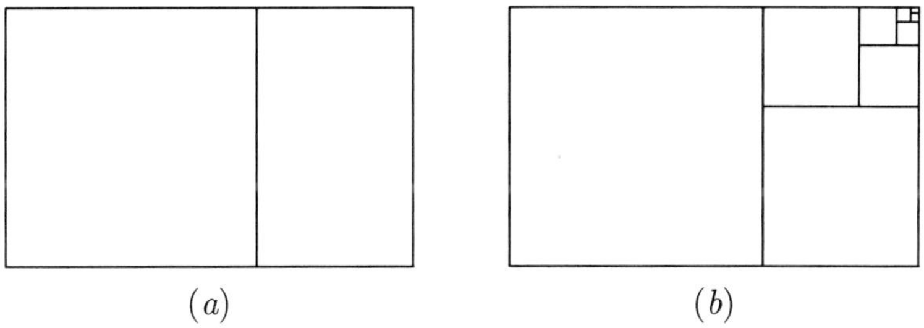

FIGURE 8.

We have already encountered probability theory, and now we approach another important branch of mathematics, the so-called *theory of dynamical systems*. The dynamical system that concerns us is a transformation of the interval $(0,1)$ to $[0,1)$, given by the following formula:

$$f : x \mapsto \frac{1}{x} - \left\lfloor \frac{1}{x} \right\rfloor = \left\{ \frac{1}{x} \right\},$$

where $\left\lfloor \frac{1}{x} \right\rfloor$ is the integer part of $\frac{1}{x}$, and $\left\{ \frac{1}{x} \right\}$ is the fractional part of $\frac{1}{x}$. (You can try to figure out how this formula is related to our problem).

Let us draw a graph of this function. It fits completely into a unit square (cf. Figure 9) . If $x = 1$, then $\frac{1}{x} = 1$ and $\left\{ \frac{1}{x} \right\} = 0$. As x begins to decrease, $\frac{1}{x}$ grows, and the fractional part will also grow, until the integer part of $\frac{1}{x}$ equals 1. When x gets down to $\frac{1}{2}$, then $\frac{1}{x}$ is equal to 2. Thus shortly before that (when x is just a bit larger than $\frac{1}{2}$), the fractional part of $\frac{1}{x}$ is close to 1. In the interval $\left(\frac{1}{2}, 1\right)$, the graph of the function is made up of a piece of the hyperbola $y = \frac{1}{x}$, moved one unit downwards. Likewise, between one-half and one-third we again get a piece of this hyperbola, only lowered by 2, and in general, on each interval $\left(\frac{1}{k+1}, \frac{1}{k}\right)$, the graph of f is a piece of a hyperbola, shifted downwards by k.

Theorem. *The transformation f has an invariant measure.*

This means the following.

We define a mass on the interval $(0,1)$. That is, we define a "density function" $\rho(x)$ and we assign to a set $A \subset (0,1)$ the mass

$$\mu(A) = \int_A \rho(x)dx.$$

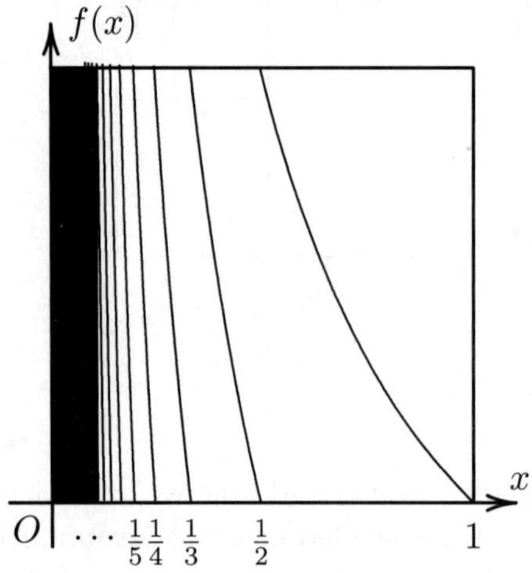

FIGURE 9.

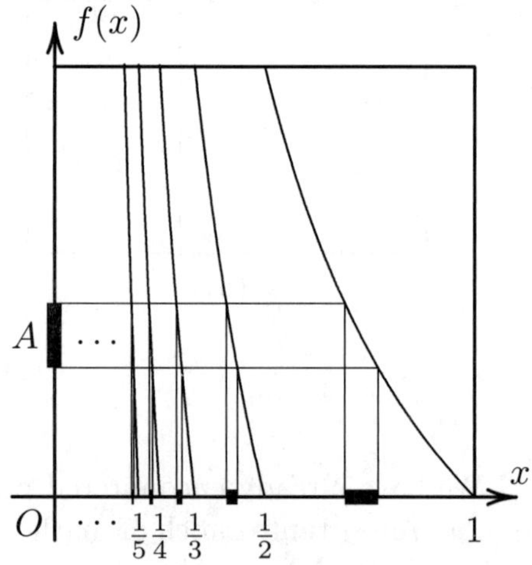

FIGURE 10.

(For A, we can take an interval.) Now take the inverse image of the interval A: this is the set of all points in $(0,1)$ that are taken to A by the map f. We denote this set by $f^{-1}A$. In our case, the inverse image consists of infinitely many pieces (cf. Figure 10). Then $\mu(f^{-1}A)$ is the sum of the measures (masses) of all these pieces. Our theorem asserts that there exists a density function ρ such that, for all intervals A,

$$\mu(A) = \mu(f^{-1}A). \tag{2}$$

This density function (although perhaps in a different form) was found by Gauss:

$$\rho(x) = \frac{1}{1+x} \cdot \frac{1}{\ln 2}$$

(the factor of $\frac{1}{\ln 2}$ is needed so that the total mass is equal to 1, as is customary in probability theory. The measure with density $\rho(x) = \frac{1}{1+x}$, without this constant, is also invariant.)

Condition (2) is equivalent to a *telescoping sum.* There is a famous problem: compute the sum

$$S = \frac{1}{1\cdot 2} + \frac{1}{2\cdot 3} + \frac{1}{3\cdot 4} + \cdots.$$

The telescopic sum works as follows. Since

$$\frac{1}{1\cdot 2} = 1 - \frac{1}{2}, \quad \frac{1}{2\cdot 3} = \frac{1}{2} - \frac{1}{3}, \quad \text{etc.,}$$

$$S = 1 - \frac{1}{2} + \frac{1}{2} - \frac{1}{3} + \frac{1}{3} - \frac{1}{4} + \cdots.$$

And now the summation proceeds "automatically": $+\frac{1}{2}$ and $-\frac{1}{2}$, $+\frac{1}{3}$ and $-\frac{1}{3}$, etc., cancel out, yielding $S = 1$.

This problem was devised for the proof of the theorem mentioned above, and it gives a hint for the proof of this theorem, which in turn leads to Kuzmin's theorem.

The point is that our system is *ergodic*. The derivative of the function f, at those points where it exists, is greater than 1 in absolute value (except at the point 1, where it equals -1). Thus a segment that is initially small will grow after an application of f, and if f is applied many times, then the original set will be "smeared" with density ρ throughout the interval $(0, 1)$.

And now, in order for the corresponding term of a continued fraction to equal k, it is necessary that the integer part be equal to k, and for this it is necessary that we are between $\frac{1}{k}$ and $\frac{1}{k+1}$. Thus the mass (measure) of the interval $\left(\frac{1}{k+1}, \frac{1}{k}\right)$ is given the value p_k.

Here we need to apply the theory of dynamical systems, but I am omitting this (because I want to discuss another theory, which also based on application of continued fractions). (See **[EC5]**). The proof of Kuzmin's theorem given in Khinchin's book uses the Ergodic Theorem of Birkhoff which was proven some time before Kuzmin and which of course Wiman could not have known. But Wiman spent 300 pages on this proof. What did he actually do? Could it be that in fact he proved Birkhoff's theorem 30 years before Birkhoff did?

Other questions related to Kuzmin's theorem which seem to me to be interesting to students are the following three conjectures, for which progress in investigation can be achieved by simple computer experiments, without any proofs at all.

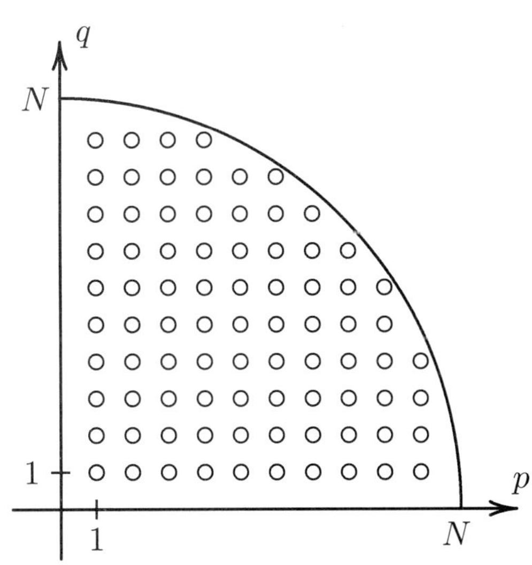

FIGURE 11.

I. Consider all the lattice points (p, q) in the positive quarter of a circle of radius N; that is, those points such that

$$p^2 + q^2 \leq N,\ p > 0, q > 0$$

(cf. Figure 11). We develop each rational number $\alpha = p/q$ into a continued fraction (all of these fractions are finite). We look at how many ones, twos, threes, etc., there are among the elements of these fractions, and determine their frequency, which will depend on N. Now let N grow very large. Will these values be close to the Gaussian probabilities from formula (1)?

On the one hand, this is an experimental question, the answer to which can be determined by computer. On the other hand, it is a theoretical question: if the computer shows that these values are close to the Gaussian distribution, the challenge is to prove this as a theorem.

II. The second question (which is close to the first, although this is not altogether obvious) is connected with a "kitchen recipe", which the rest of the world attributes to the Moscow mathematical school: the recipe of making "catsup from a cat". (In the literature I also have encountered the strange name "Arnold's Cat").

Formulating the problem, we will use the following theorem (known as Lagrange's theorem).

Theorem. *A continued fraction is periodic (that is, the sequence of its terms repeats itself starting at some point) if and only if the value represented by the fraction is a quadratic irrational number (i.e., a number of the form $a + b\sqrt{c}$, where a, b, and c are rational numbers).*

For example, the golden section, whose continued fraction has terms all equal to one, is equal to $\dfrac{\sqrt{5}+1}{2}$.

All the lattice points of the plane form a subgroup of $\mathbb{R}^2$, which is called $\mathbb{Z}^2$. An algebraist would now say that the proof of the theorem requires a consideration of the quotient group $\mathbb{R}^2/\mathbb{Z}^2$. And a geometer would now say that the plane is a universal covering of the torus (cf. Figure 12). And the two of them would be talking about the same thing. The coordinates of a point of the torus are its latitude and longitude, taken "modulo 1": we can add or subtract 1 as many times as we like to either coordinate, and obtain the same point. Therefore every point on the torus corresponds to infinitely many points on the plane.

Now let us consider the transformation A of the plane to itself which takes the point with coordinates (x, y) to the point with coordinates $(2x + y, x + y)$. More generally, we can take any transformation taking the point (x, y) onto the point $(ax + by, cx + dy)$, where a, b, c and d are integers. But now we make the assumption that the determinant $\begin{vmatrix} a & b \\ c & d \end{vmatrix}$ of this matrix is equal to 1. The transformation

$$A : (x, y) \mapsto (2x + y, x + y)$$

satisfies this condition:

$$\begin{vmatrix} 2 & 1 \\ 1 & 1 \end{vmatrix} = 2 \cdot 1 - 1 \cdot 1 = 1.$$

Note that if we add any integer to x or to y, then we are adding integers to the coordinates of the images of point (x, y). Thus, the transformation A acting on the plane $\mathbb{R}^2$ takes the lattice of integer points into itself. So we can consider it as acting on the set $\mathbb{R}^2/\mathbb{Z}^2$ (the unit square), or, equivalently, on a torus. Thus we can consider A as a transformation of a torus. More formally, the transformation A of the plane corresponds to a transformation $\widehat{A}$ of a torus.

Let the set K (the "cat") be a subset of the torus (Figure 13). First we apply $\widehat{A}$ to the "cat". The transformation $\widehat{A}$ takes each point (x, y) into some new point, which means that it takes the set K to a new set of points on the torus, which we will denote by $\widehat{A}K$. We will not draw this set on the torus itself, but on a planar "map" of the torus, as in Figure 14.[2] Applying $\widehat{A}$ again and again, we obtain sets $\widehat{A}^2K, \widehat{A}^3K$, and so on. Since the matrix for A has a determinant equal to 1, the figure $\widehat{A}K$ has the same area on the map as K itself. But its shape is completely different, and it might even turn out to be cut into separate pieces. If we keep applying $\widehat{A}$, we may get still more pieces, but the sum of the areas of the pieces will not change. After four or five applications of the transformation, the "cat" will be more or less evenly spread out along the torus, forming a sort of "sauce" (see Figure 15 (a) - (d)). This is what we mean by "making catsup of a cat".

There is a mathematical theorem which asserts that this really is a "sauce", in the following sense: if we take any region B of the torus, then the area of the intersection of the figures $\widehat{A}^k$ and B approaches the product of the areas of K and B, as k approaches infinity.[3] That is, the portion of the "cat" that lies inside B after k steps is proportional to the area of B.

Problem. Prove that for any k the transformation $\hat{A}^k$ has fixed points, and that all these fixed points for all the transformations $\widehat{A}^k$ form a dense set on the torus. We know that every transformation $\widehat{A}^k$ has a finite number of fixed points, but this number increases rapidly with k.

Let us return to continued fractions. Consider the plane which covers our torus, and the transformation A. It turns out (and is easy to prove) that there exist two lines on the plane, each of which is transformed into itself by A. One of these lines is stretched and the other squeezed, and the coefficients of stretching and squeezing must be the same since A doesn't change areas. (Note that the origin of the coordinate system is fixed: this

[2]Since the position of a point of the torus is determined by two coordinates, and we can assume that each of the coordinates belongs to the half-closed interval $[0, 1)$, points of the torus correspond to points of the square $[0, 1) \times [0, 1)$, and we can regard this square as a map of the torus.

[3]The area of the surface of the torus is equal to 1.

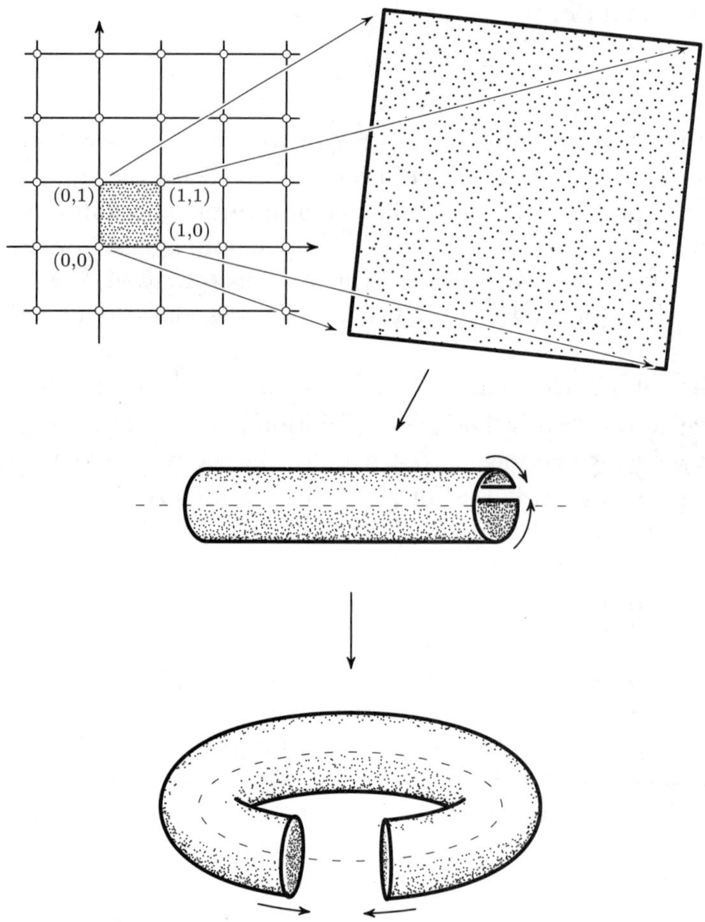

FIGURE 12.

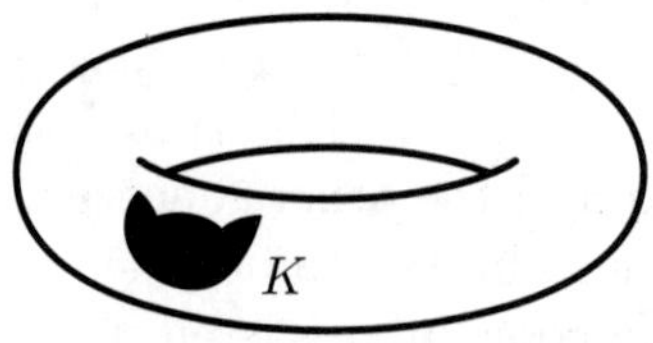

FIGURE 13.

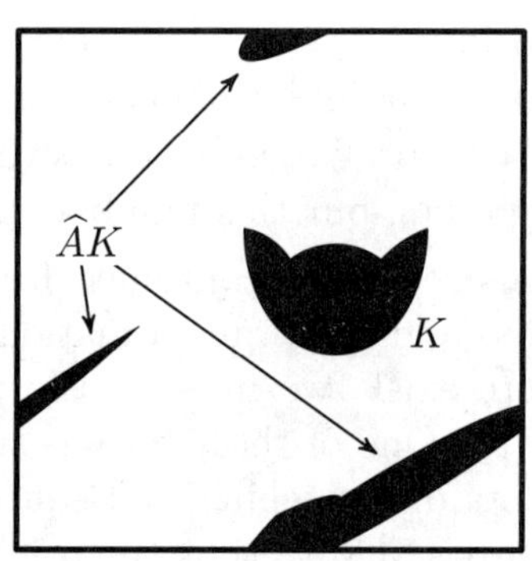

FIGURE 14.

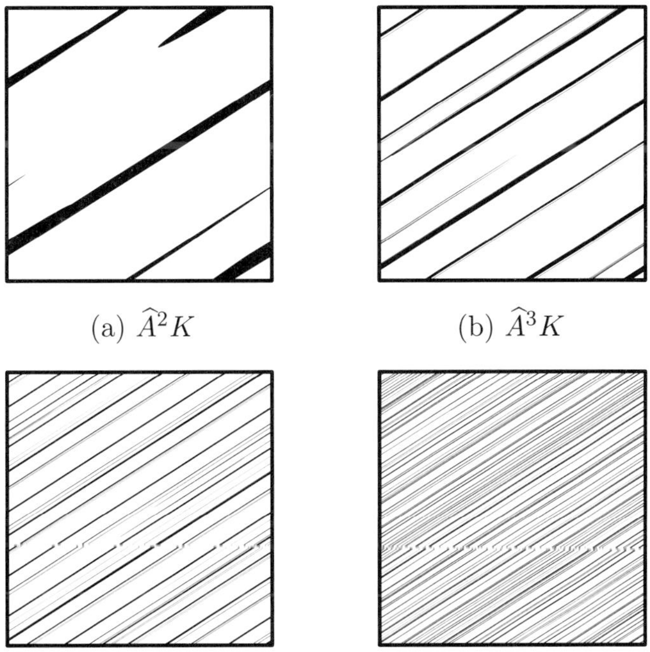

(a) $\widehat{A}^2K$ (b) $\widehat{A}^3K$

(c) $\widehat{A}^4K$ (d) $\widehat{A}^5K$

FIGURE 15.

is obvious from the formula.) A transformations such as this is called a *hyperbolic relation*, and here is why. Introduce a new system of coordinates (u, v) by taking one of these lines as the u-axis and the other as the v-axis. Then the transformation A preserves any hyperbola whose equation in this new system is of the form $uv =$ constant, for different values of the constant (Figure 16), because one of the coordinates u or v is multiplied by a certain number, while the other is divided by the same number.

It can be proven that there are no integer points on the axes Ou, Ov other than the origin. So now let us look at the set of all integer points located in one of the quadrants of the plane, and take their convex hull.[4] Transformation A takes these integer points onto integer points in the same quadrant. Therefore our convex hull is its own image, so its boundary is invariant under transformation A. It follows that the geometric characteristic of the "integer lengths"[5] of the sides of the infinite path of line segments that we constructed at the start of this article is periodic, because the operator A takes it onto itself. And these are also the terms of the continued fraction

[4]The *convex hull* of a set is the smallest convex set for which the given set is a subset.

[5]The *integer length* of a segment between two lattice points is the number of parts into which this segment is broken by lattice points. For example, the integer length of the vector (13, 21) is equal to 1 (by the theory of the golden ratio). The "probability" that an integer vector on the plane has integer length 1 (the proportion of such vectors in a

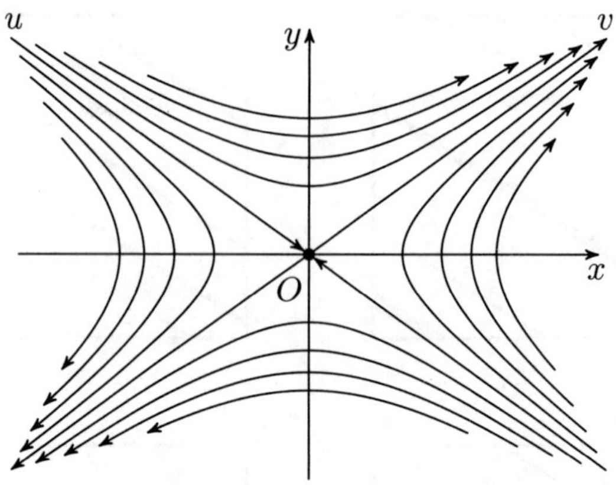

FIGURE 16.

for the number α corresponding to the line $y = \alpha x$. Strictly speaking, there are two broken line paths in Figure 1: an upper path, with vertices at e_{2s} and a lower one with vertices at e_{2s-1}. The entries of the continued fraction are the integer lengths of the segments of both broken lines, in the order $(e_1, e_3),\ (e_2, e_4),\ (e_3, e_5), \ldots$.

In fact, Lagrange's Theorem can be proven in this way. The proof given above, however, is informal, not rigorous, and does not work for every α. Also, it remains to prove the statement in the opposite direction: if the continued fraction for α is periodic, then α is a quadratic irrationality. For this, we must turn all our geometric arguments into equations, which is not hard.

Now we will formulate a second problem, which, like Problem I, requires only a computer to begin with. (Later, if the computer confirms that the hypothesis is correct, it can lead us to a non-trivial theorem.) We consider

circle whose radius approaches infinity) is equal to

$$\frac{6}{\pi^2} = \prod_p \left(1 - \frac{1}{p^2}\right) = \frac{1}{\sum_{n=1}^{\infty} n^{-2}} = \frac{1}{\zeta(2)}.$$

Here p runs through all the prime numbers, and ζ is the "zeta function",

$$\zeta(s) = \sum_{n=1}^{\infty} n^{-s} = \prod_p \frac{1}{1 - p^{-s}}.$$

The second equation applies here because of the unique factorization of a natural number n into primes. We will not give a proof here that $\zeta(2) = \dfrac{\pi^2}{6}$. It can be found in a textbook on analysis, in a discussion of the theory of Fourier series.

matrices $\begin{pmatrix} a & b \\ c & d \end{pmatrix}$, whose entries a, b, c, d are integers and whose determinant is equal to 1. We will choose from these the matrices which actually give hyperbolic rotations.[6]

There are only a finite number of matrices whose elements are not too big: what we mean here is that $a^2 + b^2 + c^2 + d^2 \leq N^2$ (for some integer N). For every such matrix there exists a line $y = \alpha x$ which gets stretched, and it is not hard to see that α is a quadratic irrationality, so its continued fraction is periodic. Let us take the period of this continued fraction and compute how many 1's there are in it, how many 2's, and so on. Then we average these numbers over all matrices $\begin{pmatrix} a & b \\ c & d \end{pmatrix}$.

In particular, we take the number of 1's in all the periodic parts of the expansions, and divide by the total number of elements in all the periods.

Conjecture. *This ratio will approach the probability given by Gauss, as N approaches infinity.*

III. We can test one more conjecture: We will do the same thing for a simple quadratic equation $x^2 + px + q = 0$, with arbitrary coefficients p and q such that the equation has real roots. To be specific, for any pairs of coefficients (p, q) which are not too big (that is, $p^2 + q^2 \leq N^2$), we will find x and develop it as a continued fraction. This continued fraction will be periodic. We take all the elements of all such continued fractions and ask if the proportion of 1's, 2's, and so on approaches Gauss' probability. This computer experiment is simpler than the previous, but the other is more interesting. In fact, neither of these conjectures have been verified.

The geometry of Lagrange's theorem: the case of arbitrary quadratic irrationalities.

If the continued fraction for a number α is periodic starting at a certain point, then α is a quadratic irrationality; that is, it satisfies a quadratic equation with integer coefficients. Indeed, for such a number α, we can write the continued fraction:

$$\alpha = a_0 + \cfrac{1}{a_1 + \ddots + \cfrac{1}{a_n + \beta}}, \qquad \beta = \cfrac{1}{b_1 + \cfrac{1}{b_2 + \ddots + \cfrac{1}{b_p + \beta}}},$$

so that α has an initial (non-repeating) segment $a_0, a_1, \ldots, a_n$ and a repeating part $b_1, \ldots, b_p; b_1, \ldots, b_p; b_1, \ldots, b_p, \ldots$.

[6] Some of these matrices give ordinary rotations. For example the matrix $\begin{pmatrix} 0 & -1 \\ 1 & 0 \end{pmatrix}$ determines the usual rotation by 90°.

We immediately obtain a quadratic equation for β, since the right hand part of the expression for β is a fractional linear function of β: $\beta = \dfrac{A\beta + B}{C\beta + D}$. For example, for $p = 1$, we have $\beta = \dfrac{1}{b_1 + \beta}$; that is, $\beta^2 + b_1\beta - 1 = 0$. Furthermore, the expression for α gives $\alpha = \dfrac{E\beta + F}{G\beta + H}$, which shows that α is a quadratic irrationality as well.

The converse statement is also true: *a continued fraction representing any quadratic irrationality α is periodic from some point on.*

We have already proven this statement geometrically for the case when the number α is the slope of a special line: the line $y = \alpha x$ which the linear transformation of the plane

$$M : (x, y) \mapsto (ax + by,\ cx + dy)$$

stretches by a factor λ. Recall that the transformation M takes the lattice $\mathbb{Z}^2$ of integer points (x, y) into itself: $M\mathbb{Z}^2 = \mathbb{Z}^2$.

The condition that the lattice of integer points be its own image under M can be expressed in terms of the coefficients a, b, c and d in the following way.

First, in order that integer points be taken into integer points, ($M\mathbb{Z}^2 \subset \mathbb{Z}^2$), it is necessary and sufficient that the coefficients be integers.

Second, in order that the image of the integer points consist of the *whole* integer lattice, and not some rarefied sublattice, it is necessary and sufficient that the image of the "fundamental parallelogram" determined by the basis vectors of the lattice ($e = (1, 0)$ and $f = (0, 1)$) be the fundamental parallelogram determined by the other two basis vectors ($E = ae + cf,\ F = be + df$). For the parallelogram determined by E and F to be a fundamental parallelogram, it is necessary and sufficient that its (oriented) area be equal to ± 1; that is, that $ad - bc = \pm 1$.

Now, let us make an explicit statement giving the numbers α for which the periodicity of the continued fraction expansion has been already proved.

Using the notation developed above, we find equations for α and λ describing the fact that the transformation M stretches the vector $e + \alpha f$ on the line $y = \alpha x$ by the factor λ in the plane $\{xe + yf\}$:

$$a + b\alpha = \lambda,\ \ c + d\alpha = \lambda\alpha.$$

Substituting the value of the coefficient of dilation λ from the first equation into the second, we get a quadratic equation for the slope α: $(a + b\alpha)\alpha = c + d\alpha$, or $b\alpha^2 + (a - d)\alpha - c = 0$. From this we obtain:

$$\alpha = \frac{d - a \pm \sqrt{(d - a)^2 + 4bc}}{2b}.$$

We are assuming that the transformation M leaves the integer lattice fixed, so the coefficients describing the images of the basis vectors in terms of the original basis vectors satisfy the relation $ad - bc = \varepsilon$ (where $\varepsilon = \pm 1$). In this case, $bc = ad - \varepsilon$, so that

$$\alpha = \frac{d - a \pm \sqrt{(d+a)^2 - 4\varepsilon}}{2b}.$$

Example: Let $a = 0$, $b = 1, d = 2p$. Then the condition that the integer lattice is fixed takes the form $c = -\varepsilon$, and we are led to the following conclusion:

Theorem. *The continued fraction of the irrational number*

(3) $$\alpha = p + \sqrt{p^2 - \varepsilon},$$

satisfying the quadratic equation

$$\alpha^2 - 2p\alpha + \varepsilon = 0,$$

where $\varepsilon = \pm 1$, *is periodic for any natural number* p.

To move towards a more general situation, let us begin with the following note. Consider the plane $\mathbb{R}^2 = \{xe + yf\}$ and the lattice Γ of integral linear combinations of the basis vectors e and f. It is clear from the nose stretching algorithm that the continued fraction for the slope α of line ℓ whose equation is $y = \alpha x$ on this plane depends not so much on the choice of basis, as on *the position of line* ℓ *with respect to the lattice* Γ.

To describe this dependence, we suppose, without loss of generality, that $\alpha > 0$. Let us look at the two angles into which line ℓ divides the positive quadrant:

$$\begin{aligned} Y_+ &: \quad y > \alpha x, \quad x > 0; \\ Y_- &: \quad y < \alpha x, \quad x > 0. \end{aligned}$$

Consider the points of the lattice that are in Y_+ (or that are in Y_-). The convex hull of this set is bordered below for Y_+ (or above, for Y_-) by an infinite broken path. The vertices of this broken path are the vectors of the nose stretching algorithm: the vertex $v_k = p_k e + q_k f$ on the broken path is followed by the vertex v_{k+2}. (On one of the broken paths all the indices k are even, and on the other they are odd.) The elements a_i of the continued fraction are the integer lengths of the segments of the broken path (see footnote on page 22). According to the nose stretching algorithm, $v_{k+2} = v_k + a_{k+1} v_{k+1}$, and the area of the parallelogram determined by v_k and v_{k+1} is $p_k q_{k+1} - q_k p_{k+1} = \pm 1$.

When we change from the basis $\{e, f\}$ to the new basis $\{e', f'\}$, and to new coefficients x', y' for the point $xe + yf = x'e' + y'f'$ we also change the equation of line ℓ from $y = \alpha x$ to the new equation $y' = \alpha' x'$ (for the same line).

We can choose the order and the signs of the basis vectors so that for points on the ray of line ℓ for which $x > 0$, we have $x' > 0$ and also so that in the angle Y_+ (where $y > \alpha x$, $x > 0$), the inequalities $y' > \alpha' x'$, $x' > 0$ (which determine the angle Y'_+) hold. Thus $Y'_+ \supset Y_+$ (cf. Figure 17 (a)).

Lemma. *Starting at a certain point, the boundaries of the convex hulls of the sets of points of lattice Γ in angles Y_+ and Y'_+ coincide.*

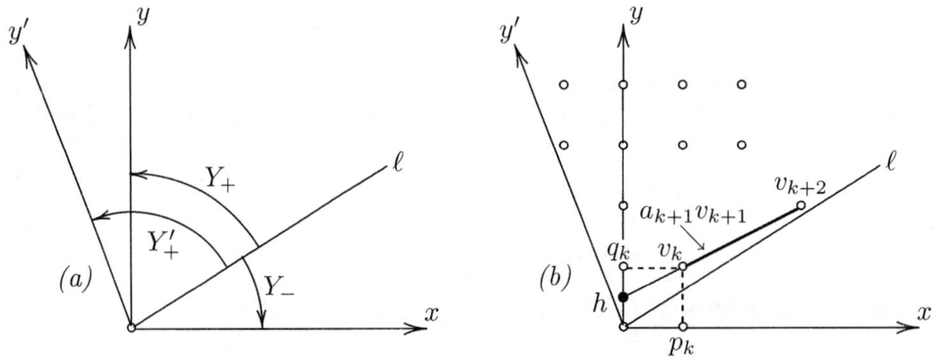

FIGURE 17.

Proof. The line connecting two neighboring vertices v_k and v_{k+2} of the convex hull lying in Y intersects the y-axis at a point whose ordinate is

$$h = q_k - p_k\frac{q_{k+1}}{p_{k+1}} = \frac{q_k p_{k+1} - p_k q_{k+1}}{p_{k+1}} = \frac{1}{p_{k+1}} \tag{4}$$

(cf. Figure 17 (b)).

Equation (4) shows that $h \leq 1$. This means that every integer point in the angle Y'_+ that is not also inside angle Y_+ lies above the line segment connecting v_k with v_{k+2}. Therefore these points have no influence on the appearance of this segment in the boundary of the convex hull of the set of integer points in the angle. Therefore the boundary of Y'_+ also contains this segment.

Of course, the boundary of the convex hull of the set of integer points of angle Y'_+ can contain additional segments, for which $x < \min\{p_k\}$ (for example, such that $x < 0$). It is only in these segments that the convex hulls differ: for sufficiently large x they are the same. This proves the lemma.

Corollary. *If the continued fraction for some number α is periodic (starting at some point) then the same is true for the number α'.*

Remark. It is easy to express α' directly in terms of α and the coefficients that express the vectors of the new basis $\{\vec{e}', \vec{f}'\}$ in terms of the old basis vectors. We obtain a fractional linear transformation

$$\alpha' = \frac{A\alpha + B}{C\alpha + D},$$

which is *unimodular*; that is, its integer coefficients satisfy the condition that they leave the area of a fundamental parallelogram fixed when passing from one basis to another: $AD - BC = \pm 1$.

In this way, whenever we have proven that the continued fraction for some number α is periodic (starting at some point) we automatically obtain the periodicity of all the numbers α' which we can obtain from α by applying a unimodular fractional linear transformation with integer coefficients.

We now show that we can eliminate the condition that the transformation changing the basis be unimodular.

Theorem. *Let the line $y = \alpha x$ be stretched by a linear transformation M of the plane which preserves the lattice $\Gamma = \mathbb{Z}^2$. Then the continued fraction of any number*

$$\alpha' = \frac{A\alpha + B}{C\alpha + D}$$

which is obtained from α by any non-degenerate integer fractional linear transformation ($AD \neq BC$) is periodic, starting at some point.

Proof. The number α' appears as a coefficient in the equation $y' = \alpha' x'$ of the line $y = \alpha x$ when it is expressed in the coordinate system generated by the pair of integer vectors $e' = Ce - Df$, $f' = -Ae + Bf$ in the plane $\{xe + yf\}$.

If the area of the parallelogram determined by these new vectors were equal to ± 1, then vectors e', f' would form a basis for the lattice Γ and everything would be proven from the arguments above. In the general case where $|AD - BC| = N > 1$, the vectors e' and f' do not generate Γ: they generate some other lattice, expanded N times, and we must amend our argument a bit.

Let Γ_0 denote the lattice generated by vectors e' and f', and let $\Gamma_1 = M(\Gamma_0)$ be the lattice generated by the vectors $M(e')$ and $M(f')$. Since the transformation M preserves areas, these new basis vectors form a fundamental parallelogram with the same area N as the parallelogram generated by vectors e' and f'. The same is true of every lattice $\Gamma_s = M^s(\Gamma_0)$ generated by a pair of vectors which determine a parallelogram of area N.

Lemma. *The sub-lattices in $\mathbb{Z}^2$ generated by a pair of vectors which determine a parallelogram of area N are finite in number (and this number is bounded by a constant depending only on N).*

Proof. A parallelogram with sides and diagonals longer than $\sqrt{N}$ would have an area larger than $N\sqrt{3} > N$. Therefore a sub lattice of this type contains a point P no farther from O than $\sqrt{N}$.

We draw line QQ' parallel to OP through point Q. In order that the area of the parallelogram determined by vectors OP and OQ be equal to N, the distance from line QQ' to OP must be equal to $\dfrac{N}{|OP|} < N$. As we iterate the transformation M, point Q of our sub-lattice that lies on this line, forms an arithmetic progression with common difference $|OP| < \sqrt{N}$. Therefore the number of different sub-lattices which we can get for various choices of point Q is not greater than $\sqrt{N}$. If we multiply this number of choices by the number of integer points at a distance less than $\sqrt{N}$ from O (and this number is no greater than CN), we will obtain an upper bound:

the number of sub-lattices is no greater than $CN^{\frac{3}{2}}$. (For example, we can use the value $C = 4$.)

Now we note that our line $y = ax$ is stretched not only by the transformation M, but also by any power M^s of this transformation.

The transformation M permutes our lattice Γ_r with its fundamental parallelogram of area N. Because there are finitely many such lattices, we can find integers $t > s$ such that $M^t\Gamma_0 = M^s\Gamma_0$. Therefore, $M^{t-s}\Gamma_0 = \Gamma_0$, so that the lattice generated by some vectors e' and f' is its own image under the transformation M^{t-s} which stretches the line $y = ax$; that is, the line $y' = a'x'$.

Therefore the continued fraction for a number α' is periodic from a certain point on, since α' is the slope of a line which is stretched by a linear transformation of the plane which fixes the lattice Γ_0, when the equation of the line is written using the basis $\{e', f'\}$.

It is clear from the theorem we've just proved that in order to prove the periodicity (starting at some point) of the continued fraction for some quadratic irrationality α' it is sufficient to express α' as the image under some fractional linear transformation

$$\alpha' = \frac{A\alpha + B}{C\alpha + D}, \qquad AD \neq BC$$

of a quadratic irrationality

$$\alpha = \frac{d - a \pm \sqrt{(d+a)^2 - 4\varepsilon}}{2b}, \qquad \varepsilon = \pm 1$$

of a special form, for which everything is already proven. But any quadratic irrationality can easily be written in the form $\dfrac{u + \sqrt{n}}{v}$ for integers u, v, n.

It is therefore sufficient to find, for each integer n that is not a perfect square, a representative of the numbers in this class with a given n that would be the slope of a line stretched by a linear transformation that fixes the lattice of integer points.

Example 1. Let $n = 2$. The number $\alpha = \sqrt{2} + 1$ satisfies the equation $\dfrac{1}{a} = \sqrt{2} - 1$; that is, $\alpha = 2 + \dfrac{1}{\alpha}$, whence

$$\alpha = 2 + \cfrac{1}{2 + \cfrac{1}{2+\ddots}}; \qquad \sqrt{2} = 1 + \cfrac{1}{2 + \cfrac{1}{2+\ddots}}.$$

Hence we have established the periodicity of the continued fraction for any irrational number of the form $\dfrac{A\sqrt{2} + B}{C\sqrt{2} + D}$.

Example 2. Let $n = 3$. For $p = 2, \varepsilon = 1$, formula (3) gives $\alpha = 2 + \sqrt{3}$. This proves the periodicity of the continued fraction for any $\alpha' = \dfrac{A\sqrt{3} + B}{C\sqrt{3} + D}$.

Example 3. Let $n = 5$. For $p = 2, \varepsilon = -1$, formula (3) gives $\alpha = 2+\sqrt{5}$. This proves the periodicity of the continued fraction for any $\alpha' = \dfrac{A\sqrt{5}+B}{C\sqrt{5}+D}$.

Example 4. For $n = 6$, $p = 5, \varepsilon = 1$, we have $\alpha = 5 + 2\sqrt{6}$. This gives the periodicity of $\alpha' = \dfrac{A\sqrt{6}+B}{C\sqrt{6}+D}$.

Example 5. For $n = 7$, $p = 8, \varepsilon = 1$, we have $\alpha = 8 + 3\sqrt{7}$. This gives the periodicity of $\alpha' = \dfrac{A\sqrt{7}+B}{C\sqrt{7}+D}$.

Example 6. For $n = 8$, $p = 3, \varepsilon = 1$, we have $\alpha = 3 + \sqrt{8}$, and the periodicity of $\alpha' = \dfrac{A\sqrt{8}+B}{C\sqrt{8}+D}$ (which could also have been obtained from an examination of the case $n = 2$).

Example 7. For $n = 10$, $p = 3, \varepsilon = -1$, we have $\alpha = 3 + \sqrt{10}$ and the periodicity of $\alpha' = \dfrac{A\sqrt{10}+B}{C\sqrt{10}+D}$.

Example 8. For $n = 11$, $p = 10, \varepsilon = 1$, we have $\alpha = 10 + 3\sqrt{11}$. This gives the periodicity of $\alpha' = \dfrac{A\sqrt{11}+B}{C\sqrt{11}+D}$.

In just the same way, in order to deal with irrationalities involving $\sqrt{n}$ we must find a non-trivial ($q \neq 0$) integer solution (p, q) of one of the two equations

$$\sqrt{p^2 - \varepsilon} = q\sqrt{n}, \quad \varepsilon = \pm 1;$$

that is, one of two equations, the first of which is called (erroneously) Pell's equation:

$$p^2 - nq^2 = 1, \quad p^2 - nq^2 = -1.$$

Theorem. *For any integer n which is not the square of another integer, Pell's equation has a non-trivial ($q \neq 0$) integer solution.*

The periodicity (from some point on) of continued fractions for all irrational numbers of the form $\dfrac{A\sqrt{n}+B}{C\sqrt{n}+D}$, with integer coefficients A, B, C, D (and $AD \neq BC$) follows from this theorem, as proven earlier.

Here are several of the simplest solutions of Pell's equation:

$$
\begin{array}{rlrl}
3^2 - 2 \cdot 2^2 = 1, & & 1^2 - 2 \cdot 1^2 = -1; \\
2^2 - 3 \cdot 1^2 = 1; & & \\
9^2 - 5 \cdot 4^2 = 1, & & 2^2 - 5 \cdot 1^2 = -1; \\
5^2 - 6 \cdot 2^2 = 1; & & \\
8^2 - 7 \cdot 3^2 = 1; & & \\
3^2 - 8 \cdot 1^2 = 1; & & \\
19^2 - 10 \cdot 6^2 = 1, & & 3^2 - 10 \cdot 1^2 = -1; \\
10^2 - 11 \cdot 3^2 = 1; & & \\
7^2 - 12 \cdot 2^2 = 1; & & \\
649^2 - 13 \cdot 180^2 = 1; & & 18^2 - 13 \cdot 5^2 = -1; \\
15^2 - 14 \cdot 4^2 = 1. & &
\end{array}
$$

Multidimensional Continued Fractions

The geometry of numbers allows us to carry over many of the constructions in the theory of continued fractions to "multi-dimensional continued fractions", in which the plane is replaced by the n-dimensional (for example, three-dimensional) space $\mathbb{R}^n$, equipped with the integer lattice $\mathbb{Z}^n$. A line is replaced with a "simplicial cone".

The integer points that are inside (but not on) the cone form a semigroup, and their convex hull is bounded by a polyhedral surface (with infinitely many faces, in general). The geometry of this polyhedral surface (called the *sail* of the original cone[7]) turns out to be the multi-dimensional generalization of the theory of continued fractions (in which the "sail" plays the role of the broken path in the nose stretching algorithm (see Figure 1).

The theory of multidimensional continued fractions is somewhat young, and I will give just a few results.

A Generalization of Lagrange's Theorem

We consider area-preserving linear transformations of $\mathbb{R}^n$ that have n invariant hyperplanes. A very simple example is the mapping in $\mathbb{R}^3$ given by the matrix

$$
\begin{pmatrix} 3 & 2 & 1 \\ 2 & 2 & 1 \\ 1 & 1 & 1 \end{pmatrix},
$$

which takes the point (x, y, z) to $(3x + 2y + z, 2x + 2y + z, x + y + z)$. We will assume that each trihedral (or n-hedral) angle into which these planes divide the space, is mapped onto itself (see Figure 18). From Dirichlet's theorem about units in algebraic number theory, it follows that the sail of each such n-hedral angle has a symmetry group generated by $n - 1$ commuting transformations, each of which preserves both the lattice of integer points and also our n-hedral angle.

[7] Translators' note: Also called the *integer hull*.

From this observation it is clear that a sail in three-dimensional space is doubly periodic (like the map of the torus): each face is infinitely repeated, like the infinitely-repeated image of the cat on the map of the torus (Figure 19). The simplest examples of such sails are described in [**29**].

Thus the two-dimensional continued fractions which correspond to cubic irrationalities are doubly periodic, although ordinary continued fractions for the same numbers (which correspond to paths wandering along this doubly periodic surface) appear chaotic and do not possess any property of periodicity.

Conversely, the generation of the cone from a linear transformation, and the connection of the sail with it algebraic "eigenvalue", whose degree is equal to that of the space, all follow from the topological periodicity of combinatoric construction of the sail. These generalizations of Lagrange's theorem (which corresponds to the case $n = 2$ and ordinary periodicity) are described in articles [**33**] and [**28**].

Tsushiyashi proved the topological periodicity of an algebraic sail. His proof is based on Dirichlet's theorem about units from algebraic number theory. His theory can also be generalized to the case of "complex eigenvalues", in which some of the invariant hyperplanes are complex.

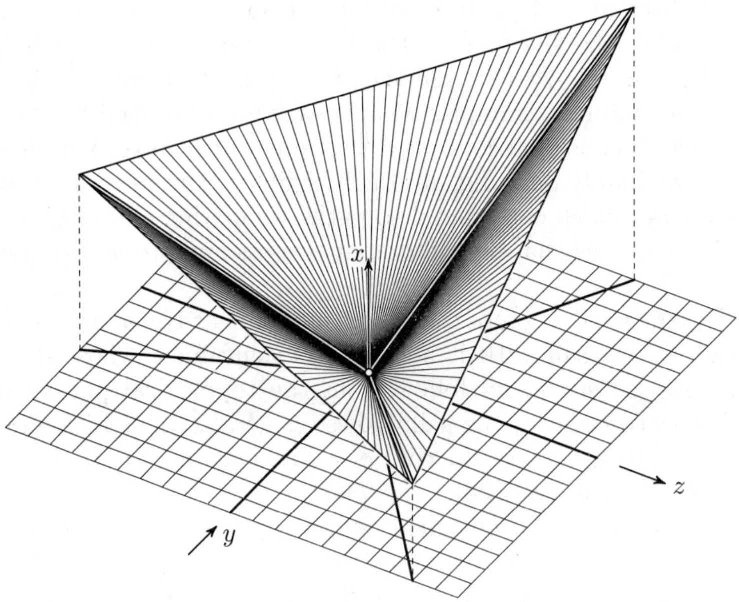

a) Trihedral angle, formed by the invariant planes of transformation (4).

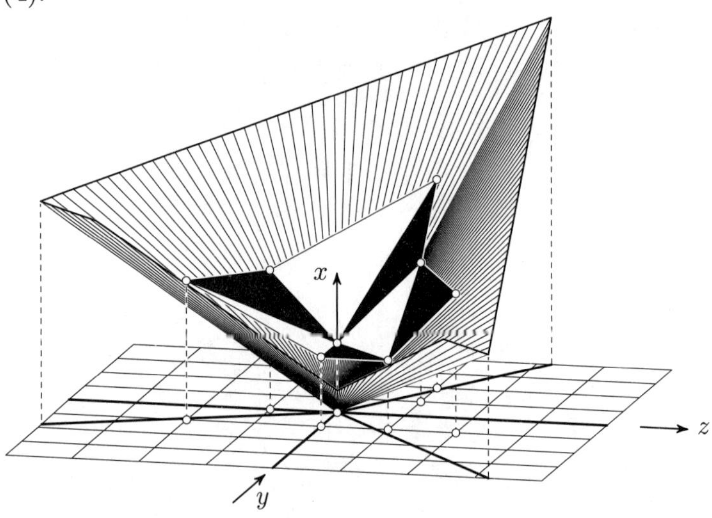

b) Fragment of the sail of the trihedral angle in figure (a), in the neighborhood of the origin.

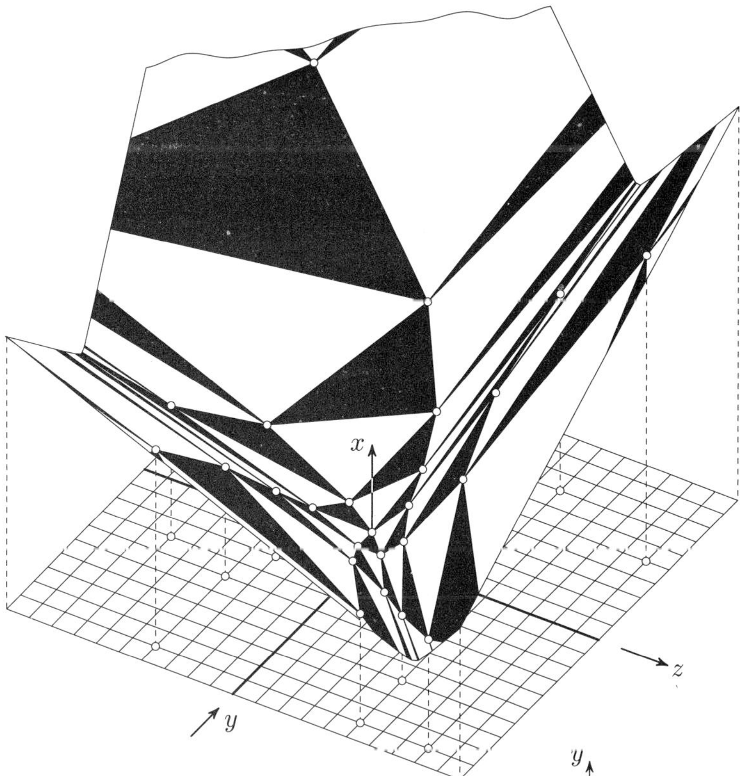

c) The larger part of the sail of figure (b).

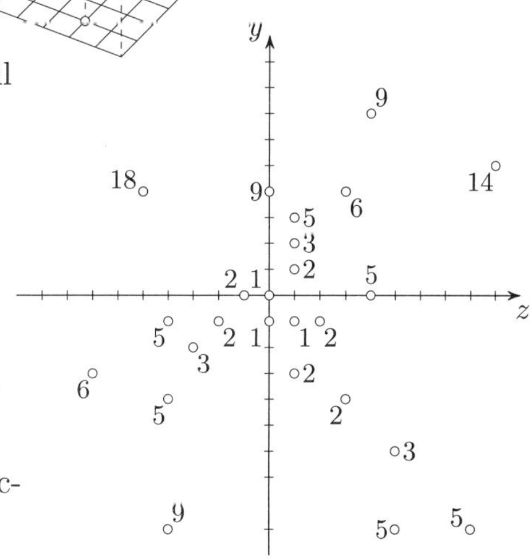

d) A projection of the vertices of the sail in figure (a) on the zy-plane along the x-axis. The x-coordinate of the corresponding vertex is provided next to each projection. (This diagram is borrowed from article [**29**].)

FIGURE 18.

a) The surface $u_1u_2u_3 = 1$ (a "generalized hyperbola"). Here u_1, u_2, u_3 are coordinates along the eigenvectors of transformation (4).

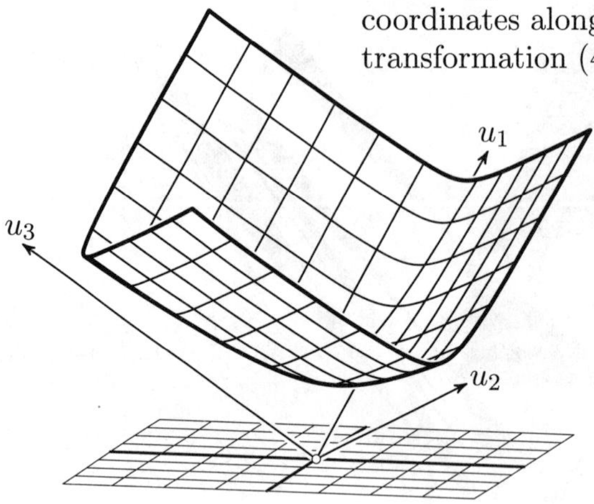

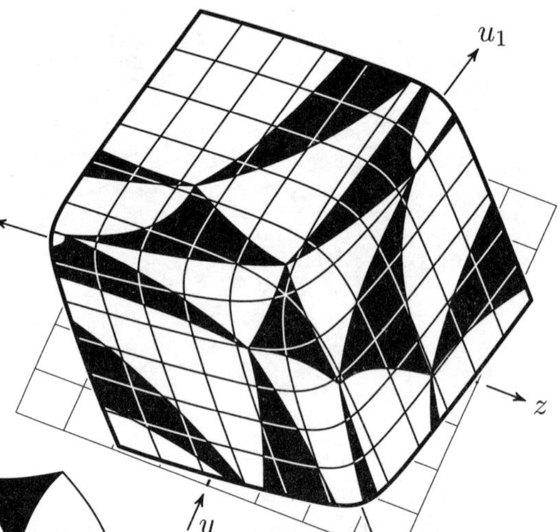

b) A central projection of the sail on the surface of figure (a).

c) A picture of the projection of the image of the surface in figures (a) and (b) under the transformation $v_t = \ln u_t$ onto the plane $v_1 + v_2 + v_3 = 0$.

FIGURE 19.

Korkina has proved the algebraic origin of a topologically periodic sail. In the case of ordinary continued fractions, this is the easier part of Lagrange's theorem. But in the case of multi-dimensional continued fractions, this is the more difficult part (and a detailed proof of Korkina's theorem has not yet been published).

In the multi-dimensional case, by the way, a basic question remains open: which triangulations of the torus, and which choices of "integer points" on the faces of such a triangulation correspond to partitions of the sail of an algebraic irrationality into convex faces? This question is open even for the two-dimensional torus and cubic irrationalities. (For one-dimensional continued fractions there is no such question: any sequence of integers can be taken as the period.)

A generalization of the statistical distribution of terms of continued fractions

I first encountered multidimensional continued fractions while attempting to classify graded commutative associative algebras (see [**5**] and [**6**]). I formulated some questions about the statistical distribution of characteristics of the sail of an arbitrary n-hedral angle in n-dimensional space: the proportion of triangular, quadrangular, etc., faces, their integral areas, the integral lengths of their edges, the number of edges at one vertex, and so on. For example, in a two-dimensional sail, will there be more triangular faces than quadrangular? Will the average number of lattice points on its edges be larger or smaller than for the edges of an arbitrary one-dimensional or three-dimensional sail? Or larger or smaller than on an arbitrary segment connecting lattice points (in spaces of various dimensions)?

As far as I know, all these questions remain open to this day. But Yu. M. Sukhov and M. L. Kontsevich, with whom I shared these questions, were able to prove that the answers to all such questions actually exist; that is, the desired statistics of the mean values over an increasing finite part of a sail are *universal* (they do not depend on the original simplicial angle) for almost all n-dimensional simplicial angles (in the sense of the Lebesgue measure). This fact allows for their description in terms of the ergodic theory of dynamical systems with $(n-1)$-dimensional time.

However, the computation of the answers (analogous to Gauss' distribution) is difficult, and is connected with the summation of series of "polylogarithms".

To achieve these results, Sukhov and Kontsevitch had to reverse my question: instead of studying the statistics of a sail of a random n-hedral angle with its vertex at the origin with respect to a fixed lattice $\mathbf{Z}^n$ in $\mathbf{R}^n$, they fix the angle; that is, a coordinate system in $\mathbf{R}^n$, for which the hyperplanes of the angle are coordinate hyperplanes. They then choose a random lattice (generated by n vectors $e_1, \dots, e_n$ which determine a parallelepiped of the unit volume).

All such ordered n-tuples of n vectors form a group $SL(n, \mathbb{R})$ of matrices of order n with real entries and determinant 1. The dimension of this smooth submanifold of the space $\mathbb{R}^{n^2}$ is equal to n^2-1. But an (ordered) n-tuple of vectors is not identical to the lattice they generate: the same lattice may be obtained from several ordered n-tuples. For example, we can replace the vector e_2 with the vector $e_1 + e_2$, and the lattice generated will not change. All such choices of basis for the lattice form a group $SL(n, \mathbb{Z})$ of integer matrices in $SL(n, \mathbb{R})$. The manifold of lattices is the quotient space $SL(n, \mathbb{R})/SL(n, \mathbb{Z})$ formed by the choice of basis vectors, considered up to a change of basis.

The theory of dynamical systems with $(n-1)$-dimensional time H is now applied to the action on the (n^2-1)-dimensional "phase space" M of the group of diagonal matrices of determinant 1 (this is the "Cartan subgroup" H of $SL(n, \mathbb{R})$). This action turns out to be ergodic (just like the transformation $x \mapsto \{1/x\}$ in the Gaussian theory). The orbit of a point under this action is smeared along M (just as our cat is spread into catsup on the surface of a torus). The desired statistical characteristics of a sail are expressed in terms of the geometry of this spread-out orbit.

Namely, consider the "diagonal vector" $(1, \ldots, 1)$ in our system of coordinates. Call a point in M (that is, a lattice) *special*, if the line determined by the diagonal vector intersects the sail corresponding to the point of M at a point belonging a face of the sail of dimension less than $n-1$ (not in general position). Special points form a hypersurface (of dimension n^2-2) in the (n^2-1)-dimensional manifold M of all lattices in n-dimensional space. The properties of the sail can be expressed in terms of the intersection of the orbit of the Cartan subgroup H with this hypersurface: the partition of the orbit into pieces, separated by the hypersurface, models the partition of the sail into its convex faces.

Unfortunately, even such properties of this hypersurface as the homology of its complement, the trace of which on the orbit determines the facets of the sail, have not been calculated.

The book [**36**] gives much more information about these theories.

Continued fractions and graded algebras

I had been attempting to invent a mathematical theory which was not related to anything, not useful for anything, and not interesting, by starting with some arbitrary axioms, like the algebraists or Bourbakists. The appearance in this work of continued fractions was completely unexpected.

A graded commutative associative algebra (over a field of real or complex numbers) is a direct sum of vector spaces of "homogeneous elements of degree d", equipped with the operation of multiplication, in which the degrees of the homogeneous factors are added (just as in the case of polynomials and their usual degrees).

Let us denote by p_n the dimension of a vector space of the homogeneous component of degree n. The series

$$\sum_{n=1}^{\infty} p_n t^n$$

is called a "Poincaré series" of an algebra (lately it has been renamed as "Hilbert series", in accordance with the Bourbaki policy of discrimination against geometry).

The Poincaré series of the algebra of polynomials of a single variable (with the usual definition of degree) is given by

$$\frac{1}{1-t} = 1 + t + t^2 + \cdots .$$

My goal was to classify the graded algebras with exactly this Poincaré series (that is, with one-dimensional spaces of homogeneous elements of any non-negative degree).

In classifying algebras with three multiplicative homogeneous generators (x, y, z) of fixed degrees $(1, u, v)$, $1 < u < v$, I discovered that there are only finitely many of them. Mathematics is an experimental science, so at first I calculated the number of algebras for small values of the degrees u and v. The result was a somewhat enigmatic table of numbers of distinct (non-isomorphic) graded algebras:

u \ v	3	4	5	6	7	8	9	10	11	12	13
2	5	1	5	1	5	1	5	1	5	1	5
3		7	7	1	7	7	1	7	7	1	7
4			9	5	9	1	9	5	9	1	9
5				11	9	9	11	1	11	9	9
6					13	7	5	13	1	13	7
7						15	11	11	11	11	15

Next I had to guess at a formula which generated the number of algebras in terms of the degrees of the generators. Using the periodicity observed in the table, I finally found that the number of algebras is related to the representation of the ratio $\frac{v}{u}$ as a continued fraction. In fact, it is equal to

$$2(a_1 + a_2 + \cdots) + 1,$$

where the a_i are the terms of the continued fraction

$$\frac{v}{u} = a_0 + \cfrac{1}{a_1 + \cfrac{1}{a_2 + \ddots}}.$$

For example, there are exactly five algebras with multiplicative generators of degree $(1, 2, 3)$, since

$$\frac{3}{2} = 1 + \frac{1}{2}, \quad a_1 = 2, \quad 2a_1 + 1 = 5.$$

In trying to classify algebras with a larger number of generators, the place of continued fractions is taken by multi-dimensional continued fractions on a polyhedral integer surface, and the problem of their classification is not yet resolved. Difficulties in laborious computation have been overcome only by powerful computer facilities to investigate the Gröbner bases (which are an effective algorithmic version of the "theological" geometry of Hilbert on the one hand, and on the other hand a contemporary computer variant of Newton's theory of polyhedra, which he considered his greatest mathematical achievement). This theory was originated in the investigation of asymptotic solutions to equations with fractional derivatives.

D. Eisenbud constructed the first examples of continual families of pairwise non-isomorphic graded commutative algebras with fixed degrees of commutative generators. Then B. Sturmfels, using computers, found more examples of sets of four such degrees, for which this is possible, in particular, the sets $(1, 3, 4, 7)$, $(1, 3, 4, 9)$, $(1, 4, 5, 6)$, $(1, 4, 5, 9)$, $(1, 5, 6, 7)$, $(1, 5, 6, 8)$, $(1, 5, 7, 8)$, $(1, 6, 7, 8)$, $(1, 6, 7, 9)$, $(1, 7, 8, 9)$ [**32**].

But a listing of all "simple" 4-tuples (for which the classification of algebras is finite) is still lacking.

My attempt to construct a useless theory turned out to be completely unsuccessful: the theory of multi-dimensional continued fractions that resulted is clearly interesting, and unites many areas of mathematics.

Editors' Comments

[EC1] Another way to prove this is to use the connection between continued fractions and the Euclidean algorithm. Let α be a positive real number; for simplicity, let us assume that it is irrational. Then we successively apply "division with remainder":

$$\begin{aligned} \alpha &= a_0 \cdot 1 + b_1 \\ 1 &= a_1 \cdot b_1 + b_2 \\ b_0 &= a_2 \cdot b_2 + b_3 \\ &\vdots \end{aligned}$$

where a_0 is a non-negative integer, $a_1, a_2, \ldots$ are positive integers, and $0 < b_i < a_i$. It is obvious that

$$\alpha = a_0 + \cfrac{1}{a_1 + \cfrac{1}{a_2 + \ddots}}.$$

We project all the vectors $\vec{e_i}$ onto a line ℓ perpendicular to the line $y = \alpha x$ and choose a linear parameter on ℓ in such a way that the image of $\vec{e_2}$ is -1. Then the images of $\vec{e_1}$, $\vec{e_2}$, $\vec{e_3}$, $\vec{e_4}, \ldots$ are α, -1, b_2, $-b_3, \ldots$. This shows that the numbers a_0, a_1, $a_2, \ldots$ from the nose stretching algorithm are the same as the numbers a_0, a_1, $a_2, \ldots$ in the continued fraction.

[EC2] Consider the vectors $\overrightarrow{OA_1}, \overrightarrow{OA_2}, \overrightarrow{OB}$ and $\overrightarrow{OA_1} + \overrightarrow{OA_2} = \overrightarrow{OA}$ (see Figure 20). We want to prove that the area of the parallelogram $OACB$ is equal to the sum of the areas of parallelograms OA_1C_1B and OA_2C_2B.

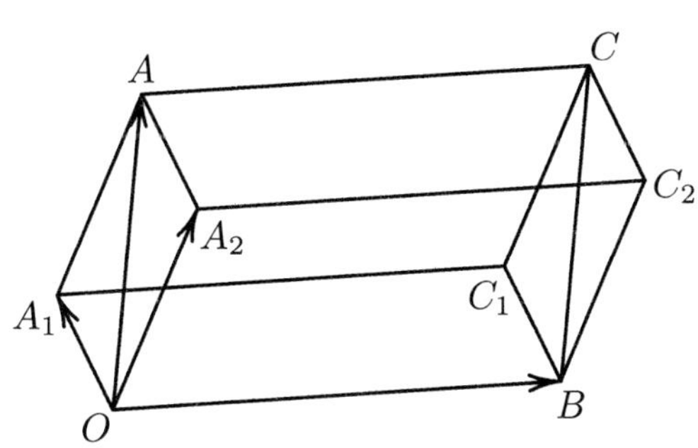

FIGURE 20.

Our drawing contains two congruent triangles, OA_1A and BC_1C, and two congruent parallelograms OA_1C_1B and A_2ACC_2. Using these, we obtain:

$$\begin{aligned}\text{area}(OACB) &= \text{area}(OACB) + \text{area}(OA_1A) + \text{area}(BC_1C)\\ &= \text{area}(OA_1ACC_1B)\\ &= \text{area}(OA_1C_1B) + \text{area}(A_1ACC_1)\\ &= \text{area}(OA_1C_1B) + \text{area}(OA_2C_2B).\end{aligned}$$

Notice that the drawing and the computations will be a bit different, if the direction of the vector $\overrightarrow{OB}$ is *between* the directions of vectors $\overrightarrow{OA_1}$ and $\overrightarrow{OA_2}$. In this case, the (signed) areas of the parallelograms OA_1C_1B and OA_2C_2B will have opposite signs, and the equality to prove will be $\text{area}(OA_1C_1B) - \text{area}(OA_2C_2B) = \pm\,\text{area}(OACB)$. The proof will be basically the same.

[EC3] We consider a function $\Delta(v_1, v_2)$ whose arguments are vectors v_1, v_2 in the plane and whose values are real numbers. We assume that this function is linear with respect to the second argument, skew-symmetric ($\Delta(v_2, v_1) = -\Delta(v_1, v_2)$), and that $\Delta(e_1, e_2) = 1$ for some orthonormal basis (e_1, e_2) in the plane. We want to prove that if $v_1 = ae_1 + be_2$ and $v_2 = ce_1 + de_2$, then $\Delta(v_1, v_2) = ad - bc$. To do this, first notice that since Δ is skew-symmetric it is also linear with respect to the first argument, and also (for an arbitrary v) that $\Delta(v, v) = 0$. Hence, we have

$$\begin{aligned}\Delta(ae_1 + be_2, ce_1 + de_2) &= ac\Delta(e_1, e_1) + ad\Delta(e_1, e_2)\\ &\qquad + bc\Delta(e_2, e_1) + bd\Delta(e_2, e_2)\\ &= ad - bc.\end{aligned}$$

[EC4] There exists a beautiful formula for the precise value of the error of the approximation of α by $\frac{p_k}{q_k}$. Namely, $\left|\alpha - \frac{p_k}{q_k}\right| = \frac{1}{\lambda_k q_k^2}$ where

$$\lambda_k = a_{k+1} + \cfrac{1}{a_{k+2} + \cfrac{1}{a_{k+3} + \ddots}} + \cfrac{1}{a_k + \cfrac{1}{a_{k-1} + \ddots + \cfrac{1}{a_1}}}.$$

It is not hard to prove this formula using the geometry of the nose stretching algorithm, but the reader who prefers to avoid doing extra work can find a proof in the book "Mathematical Omnibus" of D. Fuchs and S. Tabachnikov [**23***], Section 1.X.

In particular, it is always true that

$$a_{k+1} < \lambda_k < a_{k+1} + \frac{1}{a_{k+2}} + \frac{1}{a_k} < a_{k+1} + 2.$$

Example: for $\alpha = \pi$, $\dfrac{p_3}{q_3} = \dfrac{355}{113}$, and $a_4 = 292$. A calculator will show that $\left|\pi - \dfrac{355}{113}\right| \approx 0.267 \cdot 10^{-6}$. Hence,

$$\lambda_3 = q_3^{-2}\left|\pi - \frac{355}{113}\right|^{-1} \approx 113^{-2} \cdot (0.267 \cdot 10^{-6})^{-1} \approx 293.573.$$

Indeed, $292 < \lambda_3 < 294$.

Exercise. Deduce from our formula the following *Hurwitz-Borel Theorem*: *for every irrational number α, there exist infinitely many irreducible fractions $\dfrac{p}{q}$ such that $\left|\alpha - \dfrac{p}{q}\right| < \dfrac{1}{\sqrt{5}q^2}$.*

[EC5] Let X be a space with measure μ such that $\mu(X) = 1$, and let $f\colon X \to X$ be a measure preserving transformation (that is, if $A \subset X$ is measurable, then $f^{-1}(A)$ is measurable, and $\mu(f^{-1}(A)) = \mu(A)$). The transformation f is called *ergodic* if any measurable subset $B \subset X$ such that $f^{-1}(B) = B$ has measure 0 or 1. The main property of an ergodic transformation (called the *ergodic theorem*) is that that the "time average" equals the "space average", which means the following: For any measurable set $A \subset X$ and almost all points $x \in X$

$$\lim_{n\to\infty} \frac{\#\{k \mid 0 \le k < n, f^k(x) \in A\}}{n} = \mu(A)$$

("almost all" means that the set of those x, for which this limit does not exist or is not equal to $\mu(A)$ has measure 0). For the transformation $f\colon [0,1] \to [0,1]$, $f(x) = \left\{\dfrac{1}{x}\right\}$, its ergodicity can be deduced from the fact that f is almost everywhere differentiable and has derivative > 1; for an $\alpha \subset [0,1]$, the n-th incomplete quotient is equal to k if and only if $f^n(\alpha) \in \left[\dfrac{1}{k}, \dfrac{1}{k+1}\right)$. This argument is sufficient for completing the proof of Kuzmin's Theorem.

Part 2

Geometry of Complex Numbers, Quaternions, and Spins

Geometry of Complex Numbers, Quaternions, and Spins

The theory of complex numbers, quaternions and spins belongs to a small number of the most fundamental parts of geometry, which have important applications in physics. The French ascribe the geometric construction of the complex numbers given below to Argand, although the Danish mathematician Wessel published it seven years before Argand. (Wessel, by the way, had in mind applications of his construction to engineering; for example, to the theory of Maxwell's equations and alternating current, although this theory had not been developed yet.)

Complex Numbers

Let us consider an orthonormal system of coordinates on the Euclidean plane:

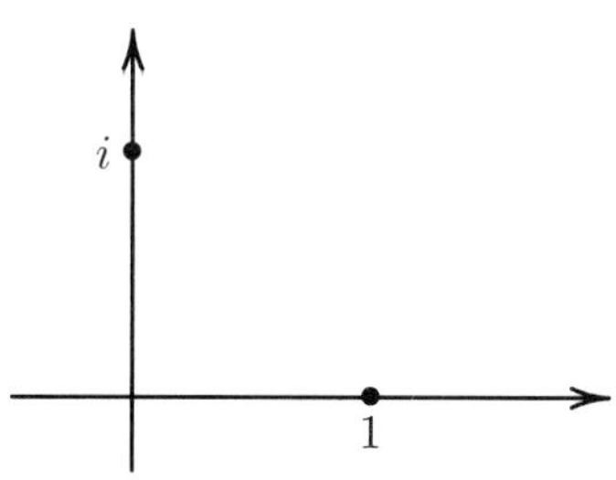

FIGURE 21. Real basis vectors in the plane of complex numbers.

We will denote the basis vectors along one axis by 1 and along the other axis by i (from the word "imaginary"). We will represent a point on the plane as $a + b \cdot i$ (without writing the 1 next to the variable a):

We can add vectors on the plane:

$$\begin{array}{rcccc} z_1 & = & a_1 & + & b_1 \cdot i \\ z_2 & = & a_2 & + & b_2 \cdot i \\ \hline z_1 + z_2 & = & (a_1 + a_2) & + & (b_1 + b_2) \cdot i \end{array}$$

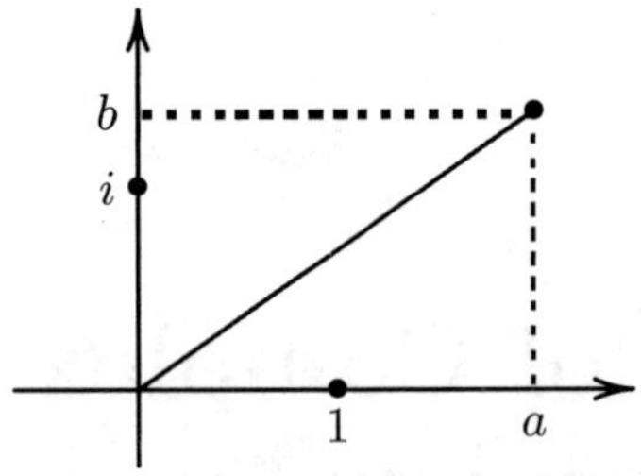

FIGURE 22. The real and complex parts of a complex number.

We can define the multiplication of complex numbers, as well as their addition. Here is a multiplication table:

$$1 \cdot 1 = 1, \qquad 1 \cdot i = i = i \cdot 1.$$

The important work begins when we multiply i by i:

$$i \cdot i = -1.$$

We call the number i *imaginary*, because there does not exist a real number a which satisfies the equation

$$a^2 = -1.$$

The product of two arbitrary complex numbers, $z_1 = a_1 + b_1 \cdot i$ and $z_2 = a_2 + b_2 \cdot i$ is defined by the distributive law:

$$z_1 z_2 = (a_1 a_2 - b_1 b_2) + (a_1 b_2 + a_2 b_1)i.$$

In other words, we obtain the real part of the product by taking the difference of the products of a_1, a_2 and b_1, b_2. We get the imaginary part of the product by add the products a_1, b_2 and a_2, b_1. This is all there is to the definition of multiplication.

Remark. All the usual properties of multiplication (commutativity: $z_1 z_2 = z_2 z_1$; associativity: $(z_1 z_2) z_3 = z_1 (z_2 z_3)$; and distributivity with respect to addition: $z_1(z_2 + z_3) = z_1 z_2 + z_1 z_3$) clearly hold.

From the algebraic point of view, this exhausts the entire theory of complex numbers.

Motions of the Plane

Complex numbers provide a mathematical tool for describing motions of the plane. To convince the reader of this, we introduce an additional

Definition. The complex number

$$\overline{z} = a - b \cdot i$$

is called the *complex conjugate* of the number $z = a + b \cdot i$.

Geometrically, we get from z to $\overline{z}$ by reflection in the axis $O1$.

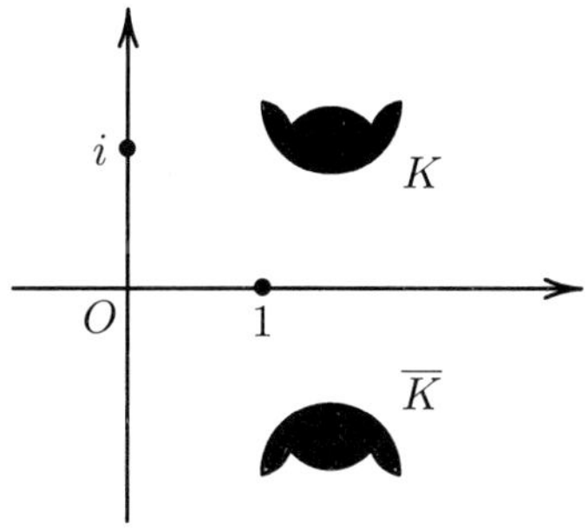

FIGURE 23. The complex conjugate of the cat K.

Theorem. *The conjugate of the sum of two complex numbers is the sum of their conjugates:*

$$\overline{z_1 + z_2} = \overline{z_1} + \overline{z_2}.$$

Theorem. *The conjugate of the product of two complex numbers is the product of their conjugates:*

$$\overline{z_1 z_2} = \overline{z_1}\,\overline{z_2}.$$

Definition. The product of a complex number with its conjugate is called the *squared modulus* of the number:

$$|z|^2 = z\overline{z}.$$

The modulus of a complex number is a non-negative real number. Indeed, the squared modulus of a complex number is not changed by conjugation:

$$\overline{z\overline{z}} = \overline{z}z = z\overline{z},$$

so it must be a real number. Also, $|z|^2 = z\overline{z} = a^2 + b^2 \geq 0$, and we take for the modulus of a complex number z the square root of $a^2 + b^2$ which is also a non-negative real number.

Remark. For complex numbers z_1 and z_2, the modulus $|z_1 - z_2|$ is equal to the distance between z_1 and z_2 in the plane of complex numbers. Indeed, if $z_1 = a_1 + b_1 \cdot i$ and $z_2 = a_2 + b_2 \cdot i$, then $z_1 - z_2 = (a_1 - a_2) + (b_1 - b_2) \cdot i$ and

$$|z_1 - z_2|^2 = (a_1 - a_2)^2 + (b_1 - b_2)^2$$

which is the square of the distance between the points (a_1, b_1) and (a_2, b_2).

Definition. The *argument* α of a non-zero complex number is equal to the angle of rotation from the positive semiaxis $O1$ in the direction of the positive (imaginary) semiaxis Oi to the direction from O to our complex number.

Remark. If $|z| = 1$, then

$$a = \cos\alpha, b = \sin\alpha.$$

Readers who are not familiar with the sine and cosine functions can take this remark for their definition.

We now use complex numbers to study motions of the Euclidean plane. Consider the complex number $w \in \mathbb{C}$. We look at the transformation of "multiplying by the complex number z", which takes each point w to the point zw, where $|z| = 1, \arg z = \alpha$ ($\arg z$ means the argument of z).

Theorem. *The transformation of multiplying by the complex number z with unit modulus is simply a rotation of the plane $\{w\}$.*

Proof. We take some complex number w, and compute the modulus of the complex number which is its image under our transformation:

$$|zw|^2 = zw\overline{zw} = (z\overline{z}) \cdot (w\overline{w}) = w\overline{w} = |w|^2.$$

It follows that any vector is taken into a vector of the same length. Also, the distance between the endpoints of two vectors remains the same:

$$|zw_1 - zw_2| = |z(w_1 - w_2)| = |w_1 - w_2|.$$

Thus, the transformation of multiplication by a complex number with unit modulus leaves lengths invariant.

An important detail: This transformation preserves orientation.

Exercise. Our transformation takes a clockwise rotation of the plane $\{w\}$ into a clockwise rotation (that is, in the same direction as the original rotation).

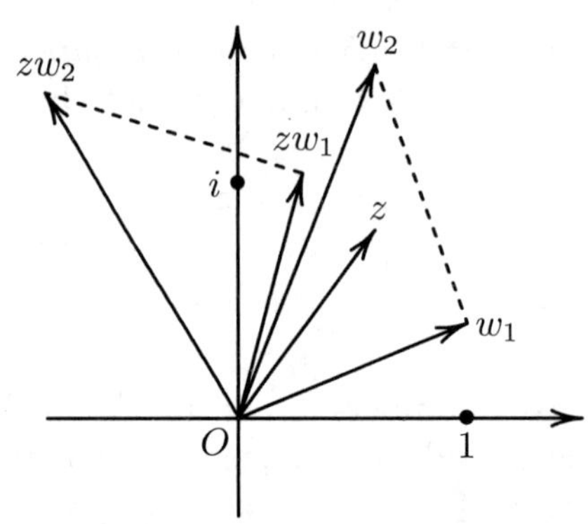

FIGURE 24. The operation of multiplication by a complex number with unit modulus.

A Digression Concerning Orientations

For a definition of orientation, we need a formula which is often hidden from the students: the formula for the area of a parallelogram.

Suppose we have parallelogram on a Euclidean plane with orthonormal coordinates $\{X, Y\}$. Denote the first vector defining the parallelogram bb $A = (x_1, y_1)$, and the second vector by $B = (x_2, y_2)$.

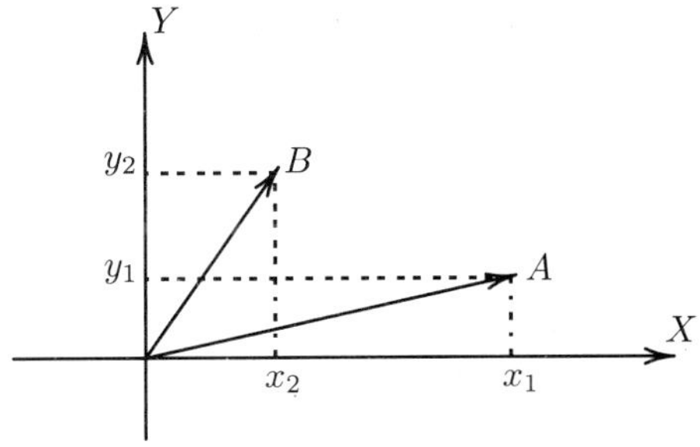

FIGURE 25. The orientation of a plane of a pair of vectors.

Theorem. *The area $S(A,B)$ of the parallelogram generated by vectors A, B is a linear function of vector A:*

$$S(A_1 + A_2, B) = S(A_1, B) + S(A_2, B).$$

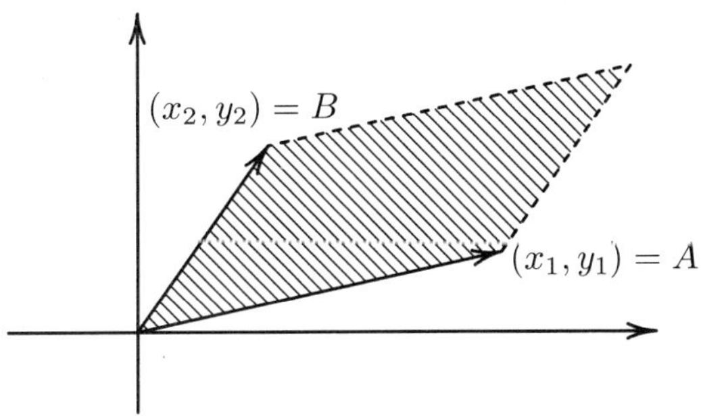

FIGURE 26. The parallelogram generated by a pair of vectors A, B.

(See [**EC1**].)

This linearity holds, if we take for $S(A,B)$ the *signed area*. Namely, we assume that the area of the parallelogram is *positive*, if the direction of rotation from A to B is the same as the direction of rotation from the first positive half-axis to the second (in Figure 6, this direction is counterclockwise). Similarly, the area of the parallelogram is *negative* if the direction of rotation from A to B is opposite. The linearity of the dependence of the area on the first vector also means that

$$S(kA, B) = kS(A, B).$$

These two simple facts conceal within them the entire "theory of determinants".

Let us take the basis $e = (1, 0)$, $f = (0, 1)$. Then we can write our vectors A and B in the form

$$A = x_1 e + y_1 f,\ B = x_2 e + y_2 f.$$

Let us compute the area $S(A, B)$. Linearity lets us write the area as the sum of four addends:

$$S(A, B) = x_1x_2S(e, e) + x_1y_2S(e, f) + y_1x_2S(f, e) + y_1y_2S(f, f).$$

Here $S(e, e) = 0$, since the parallelogram defined by the pair (e, e) of vectors is degenerate. For the same reason, $S(f, f) = 0$. We can further note that $S(e, f) = 1$. But $S(f, e) = -1$, since the direction of rotation from f to e is in the opposite sense. Therefore we have the following expression for the area:

$$S(A, B) = x_1y_2 - x_2y_1.$$

This number is called the *determinant* of the square table of the four components of our vectors. The table itself is called the *matrix* of the parallelogram:

$$S(A, B) = \begin{vmatrix} x_1 & y_1 \\ x_2 & y_2 \end{vmatrix}.$$

We now consider the question: does multiplication by z preserve orientation? We must take some fundamental parallelogram on the plane $\{w\}$ and observe its image. If the area of the image is positive, then orientation is preserved. If it is negative, then the orientation of a fundamental parallelogram (and thus of any parallelogram) will change. The images of the vectors e and f are

$$ze = z = a + bi, \qquad zf = zi = -b + ai.$$

So the matrix of the image of the parallelogram has the form

$$\begin{pmatrix} x_1 & y_1 \\ x_2 & y_2 \end{pmatrix} = \begin{pmatrix} a & b \\ -b & a \end{pmatrix}.$$

The determinant of this matrix is positive (for $z \neq 0$), since

$$\begin{vmatrix} a & b \\ -b & a \end{vmatrix} = a^2 - (-b) \cdot b = a^2 + b^2 > 0.$$

It follows that multiplication by a non-zero complex number z preserves the orientation of the plane of the images under this multiplication.

Next we ask: What is the angle through which the vector w is rotated when multiplied by z? Take the simplest vector and compute by what angle it is turned by our multiplication.

The algebraists believe that the simplest number is 0, but it does not suit us. We will take for the simplest number the number $w = 1$. The multiplication by z will take it into $z \cdot w = z = a + bi$.

It is seen from this diagram that our transformation rotates the vector $w = 1$ through an angle $\alpha = \arg z$.

Corollary. *Multiplication by a complex number z such that $|z| = 1$ is in fact a rotation of the plane through an angle equal to $\arg z$.*

Theorem. *If we multiply two complex numbers, we add their arguments:*

$$\arg(zw) = \arg z + \arg w.$$

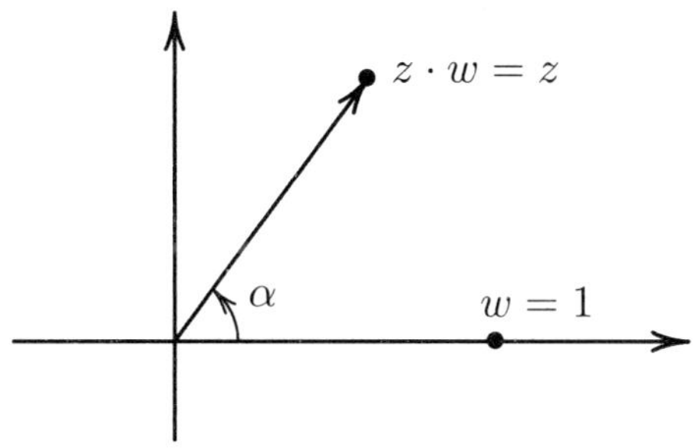

FIGURE 27. The rotation of vector w upon multiplication by the number z.

Proof. We have already shown that the vector $w = 1$ is rotated through an angle α when we multiply by the number z. But since multiplication by z is a rotation, we know that any vector w is rotated through the same angle. Let $\arg w - \beta$. Then the number zw will have an argument equal to $\alpha + \beta$, that is, $\arg(zw) = \alpha + \arg w = \arg z + \arg w$.

This leads us to certain trigonometric identities, which otherwise would be impossible to understand. Let $z = a + bi, w = c + di, |z| = 1, |w| = 1$. Then

$$zw = (ac - bd) + (ad + bc)i.$$

Since $\arg z = \alpha$, we have $a = \cos\alpha, b = \sin\alpha$. Similarly, since $\arg w = \beta$, we have $c = \cos\beta, d = \sin\beta$. Taking real and imaginary parts of the product zw, we then have:

$$\begin{aligned} \cos(\alpha + \beta) &= ac - bd = \cos\alpha\cos\beta - \sin\alpha\sin\beta \\ \sin(\alpha + \beta) &= ad + bc = \cos\alpha\sin\beta + \sin\alpha\cos\beta \end{aligned}$$

The formula for the multiplication of two complex numbers is easy to remember and difficult to confuse with anything else. It holds within itself many trigonometric identities, which makes them much beloved as questions offered to the students at various examinations.

The formulas above completely exhaust the theory of rotations and rigid motions of the Euclidean plane.

Exercise. Show that

$$\cos n\varphi + i\sin n\varphi = (\cos\varphi + i\sin\varphi)^n,$$

for any natural number n. (This formula is called *De Moivre's Theorem*, although it was discovered much earlier by a completely different person.)

Example. From this formula it follows (by expansion of the binomial on the right-hand side) that both $\cos n\varphi$ and $\sin n\varphi$ are polynomials with integer coefficients in the variables $x = \cos\varphi, y = \sin\varphi$. Or, in other words, *trigonometric polynomials* (linear combinations of sines and cosines of multiples of an angle) *can be viewed as a restriction of ordinary polynomials in two variables (x and y) to the circle ($x^2 + y^2 = 1$).*

In particular, the formulas

$$\cos(2\varphi) = \cos^2\varphi - \sin^2\varphi, \qquad \sin(2\varphi) = 2\sin\varphi\cos\varphi,$$
$$\cos(3\varphi) = 4\cos^3\varphi - 3\cos\varphi, \qquad \sin(3\varphi) = 3\sin\varphi - 4\sin^3\varphi$$

are very often useful.

Since the cosine is an even function (while the sine is an odd function) of φ, the function $\cos(n\varphi)$ can be written as a polynomial in only one variable $x = \cos\varphi$ (by replacing each occurrence of $\sin^2\varphi$ with $1-\cos^2\varphi$). These wonderful polynomials of a single variable are called the Chebyshev polynomials, and they have many useful properties ("The least deviation from zero", the phenomenon of "Lissajous figures" on the screen of an oscilloscope).[8] The simplest of these polynomials can be rewritten in the following form:

$$\mathcal{F}_1(x) = x, \mathcal{F}_2(x) = 2x^2 - 1, \mathcal{F}_3(x) = 4x^3 - 3x, \ldots.$$

However, the following question arose very early: Aside from rotations and rigid motions in the plane $\mathbb{R}^2$, there are also rotations and rigid motions in space $\mathbb{R}^3$. How do we describe these? Wessel did something about this as early as 1820. But it was the Irish mathematician Hamilton who created a complete theory somewhat later.

One dark evening, on the way home from Trinity College in Dublin Hamilton sought the aid of alcoholic spirits. This led him to a wonderful theorem, which we describe below (and which he had tried to formulate many times earlier). They say that he was so struck by the formula he had discovered, that he immediately took a pen-knife and carved it into the wooden railing on the little bridge he happened to be crossing over a canal. But although I tried to find this inscription, done by Hamilton, I could not find it on that little bridge.[9]

The Generalization of Complex Numbers to the Concept of Quaternions

It turns out that in order to describe rotations in $\mathbb{R}^3$, four numbers are necessary. So we need vectors in a 4-dimensional real space, also called *quaternions*. The formula carved into the railing by Hamilton ($ijk = -1$) gives the multiplication table for quaternions.

Thus a quaternion is a vector in four-dimensional real space, with a basis of $1, i, j, k$ (these vectors are called *basis quaternions*): $a + bi + cj + dk$. The number a is called the *real part* (or scalar part), and the three-dimensional

[8] See, for example, the book [**4**].

[9] The episode described here occurred on the 16 of October 1843 on Brookham Bridge over the Royal Canal, where Hamilton also carved the formula $ijk = -1$. In a letter to his son Archibald on August 5, 1865, Hamilton wrote of the fate of this formula: "...but, of course, as for the inscription, it has long since mouldered away" (see [**31**], pages 103–104). Now, this formula on the bridge is carved in stone.

vector $v = bi + cj + dk$ is called its *imaginary part*. The word "vector" first appeared exactly in this theory. There were no vectors before Hamilton's time, so he had to invent all the terminology.

We add two quaternions by adding their components. Addition of quaternions is commutative and associative.

The most difficult part of dealing with quaternions is figuring out how to multiply them. Multiplication must be distributive with respect to addition, so it suffices to know how to multiply the basis quaternions.

Here is Hamilton's multiplication table for the basis quaternions:

$$\begin{array}{rclrclrclrcl}
1\cdot 1 &=& 1, & 1\cdot i &=& i, & 1\cdot j &=& j, & 1\cdot k &=& k \\
i\cdot 1 &=& i, & i\cdot i &=& -1, & i\cdot j &=& k, & i\cdot k &=& -j; \\
j\cdot 1 &=& i, & j\cdot i &=& -k, & j\cdot j &=& -1, & j\cdot k &=& i; \\
k\cdot 1 &=& i, & k\cdot i &=& j, & k\cdot j &=& -i, & k\cdot k &=& -1.
\end{array}$$

Notice that two different imaginary quaternion "units" do not commute in multiplication; rather, they are "anticommutative": $ij = k$, but $ji = -k$. Once we know that $ij = k$, the rest is easily deduced from the condition that multiplication is associative. For example, $ik = iij = -j$, since $i^2 = -1$.

The rules for multiplying basis quaternions can be obtained from the rule $ij = k$ by cyclic permutation of the variables:

$$ij = k, jk = i, ki = j.$$

While multiplication is not commutative, it is associative. It has to be non-commutative to describe rotations of three-dimensional space, since the latter do not always commute.

The following identities hold for any quaternions p, q, r:

$$\begin{array}{rl}
p + q = q + p & \text{Commutativity of Addition} \\
(p + q) + r = p + (q + r) & \text{Associativity of Addition} \\
(p \cdot q) + r = p + (q \cdot r) & \text{Associativity of Multiplication} \\
p \cdot (q + r) = p \cdot q + p \cdot r & \text{Distributivity}
\end{array}$$

This operation of multiplication does not require mental effort. It is similar to the multiplication of numbers with many digits.[10]

Suppose we are given two quaternions:

$$p = a_1 + v_1; \qquad q = a_2 + v_2$$

(where $v_1 = b_1\, i + c_1\, j + d_1\, k$ and $v_2 = b_2\, i + c_2\, j + d_2\, k$). Let us compute their product. We start with the real part of the result:

$$\mathrm{Re}(pq) = a_1a_2 - b_1b_2 - c_1c_2 - d_1d_2.$$

[10]Descartes proposed a method for the complete exclusion of both geometry and imagination from mathematics, as a sort of "democratization": that is, any dullard could work with such a method just as successfully as the greatest genius.

It is clear that the product of the real parts a_1a_2 must appear in the result. But in cross multiplying other terms, which involve other basis quaternions, we get real numbers from $i^2 = -1, j^2 = -1, k^2 = -1$. And no other cross products will involve real numbers. By definition, the *scalar product* (v_1, v_2) of two vectors $v_1 = b_1\, i + c_1\, j + d_1\, k$ and $v_2 = b_2\, i + c_2\, j + d_2\, k$ in three-dimensional Euclidean space with an orthonormal basis (i, j, k) (also called the *dot-product* and denoted as $v_1 \cdot v_2$) is the following bilinear function of these two vectors

$$(v_1, v_2) = b_1b_2 + c_1c_2 + d_1d_2.$$

Exercise. Prove that the scalar product of two vectors is equal to the product of their lengths and the cosine of the angle between them:

$$(v_1, v_2) = \|v_1\| \cdot \|v_2\| \cdot \cos\angle(v_1, v_2).$$

Let us now compute the imaginary part of the product of the quarter ions pq.

$$\operatorname{Im}(pq) = a_1v_2 + a_2v_1 + [v_1, v_2],$$

where $[v_1, v_2]$ is the *vector product* of vectors v_1 and v_2 (also called the *cross-product* and denoted as $v_1 \times v_2$). Unlike the scalar product (which is a number), the vector product of two vectors is a vector. There are easy memorized formulas for the components of this vector in terms of determinants: the vector product of $v_1 = b_1\, i + c_1\, j + d_1\, k$ and $v_2 = b_2\, i + c_2\, j + d_2\, k$ is

$$[v_1, v_2] = \begin{vmatrix} c_1 & d_1 \\ c_2 & d_2 \end{vmatrix} i + \begin{vmatrix} d_1 & b_1 \\ d_2 & b_2 \end{vmatrix} j + \begin{vmatrix} b_1 & c_1 \\ b_2 & c_2 \end{vmatrix} k.$$

Exercise. Show that $[v_1, v_2]$ is perpendicular to v_1 and v_2, and that its length is equal to $\|v_1\| \cdot \|v_2\| \cdot \sin\angle(v_1, v_2)$.

The direction along the perpendicular is chosen according to the following requirement, involving orientations (known as the "right hand rule"): the triple $(v_1, v_2, [v_1, v_2])$ of vectors determines the same orientation of space as the triple (i, j, k). This means that we can continuously deform one of these triples into the other, keeping the vectors linearly independent during the deformation. For example, the triples (i, j, k) and (j, k, i) orient space in the same way, while the triple (i, k, j) orient it differently.

Exercise. The vector product of two vectors changes sign when the two vectors are interchanged.

Example. $[i, j] = k = -[j, i]; [i, i] = 0$.

Exercise. The vector product of two vectors is linear in both of them.

This description of vector multiplication completes our description of Hamilton's algebra of quaternions. The basic significance of this operation lies in its providing a description of rotations in three dimensional Euclidean space. Recall that we can describe a rotation of the oriented Euclidean plane $\mathbb{R}^2$ through an angle α is by identifying the plane with $\mathbb{C}$. The rotation is described by the transformation of multiplication by z, $w \mapsto z \cdot w$ where $z = \cos\alpha + i\sin\alpha$.

For quaternions, we must guess at an analogy to the number z. A complete description of this choice leads to the theory of spin. (It was probably Rodrigues who first discovered the formulas for this.)

We begin with Euclidean space $\mathbb{R}^3$, oriented by an orthonormal basis i, j, k. A rotation of $\mathbb{R}^3$ is defined by its oriented axis of rotation (passing through the origin) and the angle of rotation about this axis. (An orientation of the axis is needed to distinguish between clockwise and counterclockwise rotations: the rotation must appear counterclockwise to an observer located at a distant point at the positive semiaxis.) The oriented axis can be given by a unit vector v, which in turn is given by the angles α, β, γ it makes with the axes i, j, k. A rotation about this axis through a positive angle is the same as a rotation through the opposite angle around the oppositely oriented axis.

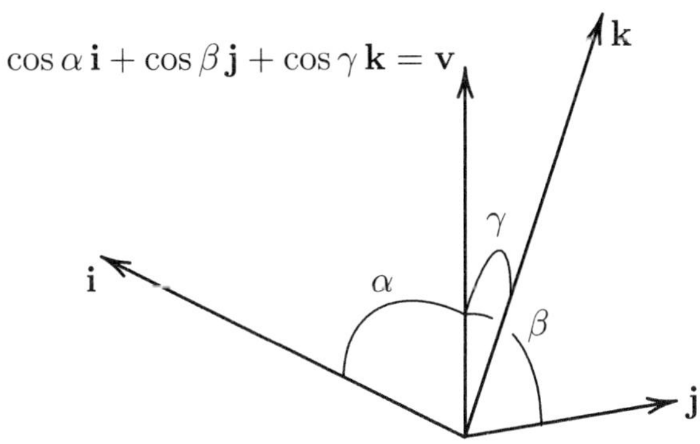

FIGURE 28. The unit vector on the axis of rotation and its directing cosines.

Let us look at the unit coordinate vector of the axis of rotation. This vector has components which are the cosines of the direction angles (formed by this vector and vectors directed along the axes). By the Pythagorean theorem, the vector v is of unit length.

We now turn to the crucial formula for quaternions, which describes rotation. This formula is a secret kept from from students in mathematics and physics (in the solid-state physics, it is revealed in the form of the Pauli matrices).

What is the dimension of the group of rotations of Euclidean three space? We use the three directing cosines and the angle of rotation to specify a rotation: four real numbers. So it would seem that the dimension of the set of rotations is four, but this is not correct. There is a relation between these four numbers: $||v|| = 1$. Thus, the dimension of the group SO(3) of rotations of Euclidean space $\mathbb{R}^3$ around a point O is equal to three.

Exercise. Compute the dimension of the group $SO(n)$ of all rotations of the spaces $\mathbb{R}^4, \mathbb{R}^5$ and so on, about the origin.

In the notation $SO(n)$, the letter O stands for orthogonality, that is, for the conservation by the transformations in the group of the lengths of vectors in $\mathbb{R}^n$; the letter S ("special") stands for the conservation of the orientation.

We now assign a quaternion of unit norm to every rotation of three dimensional oriented Euclidean space.

Definition. The quaternion $\overline{q} = a - v$ is called the *conjugate* of the quaternion $q = a + v$.

Remark. For any two quaternions q_1 and q_2, $\overline{q_1 q_2} = \overline{q}_2 \, \overline{q}_1$. Indeed, if $q_1 = a_1 + v_1$, $q_2 = a_2 + v_2$, then $\overline{q}_1 = a_1 - v_1$, $\overline{q}_2 = a_2 - v_2$ and

$$\begin{aligned}
\overline{q}_1 \, \overline{q}_2 &= (a_1 a_2 - (-v_1, -v_2)) + (a_1(-v_2) + a_2(-v_1) + [-v_1, -v_2]) \\
&= (a_1 a_2 - (v_1, v_2)) + (-a_1 v_2 - a_2 v_1 + [v_1, v_2]), \\
q_2 q_1 &= (a_2 a_1 - (v_2, v_1)) + (a_2 v_1 + a_1 v_2 + [v_2, v_1]) \\
&= (a_1 a_2 - (v_1, v_2)) + (a_1 v_2 + a_2 v_1 - [v_1, v_2]),
\end{aligned}$$

from which $\overline{q_1 q_2} = \overline{q}_2 \, \overline{q}_1$.

Definition. The *squared norm* of a quaternion is the number $q \cdot \overline{q} = a^2 - (v, (-v)) = a^2 + ||v||^2 \geq 0$.

The squared norm is always a real number, and is positive whenever $q \neq 0$. Indeed, the formula given in its definition is a sum of squares of real numbers:

$$q\overline{q} = (a + bi + cj + dk)(a - bi - cj - dk) = a^2 + b^2 + c^2 + d^2 \geq 0.$$

Thus the norm $||q|| = \sqrt{q\overline{q}}$ is a real, non-negative number, and is positive if $q \neq 0$.

Suppose we are given some quaternion with unit norm:

$$q = a + v; \quad ||q|| = 1.$$

From the definition of the norm, we have:

$$a^2 + ||v||^2 = 1.$$

It is natural to consider a and $||v||$ as the cosine and sine of some angle:

$$a = \cos\varphi, \quad ||v|| = \sin\varphi.$$

We can write our quaternion in the form $q = \cos\varphi + \sin\varphi \cdot v'$, where v' is a unit vector.

If we want to use our quaternion to describe a rotation in three-dimensional space, it is natural to take for v' the unit vector of the (oriented) axis of rotation. To define our quaternion, it remains only to choose the angle φ in the last formula. Let θ be the actual value of the angle of rotation. Contemporary physics differs from the physics of two centuries ago. For this reason, we need "Rodrigues' Halving": it turns out that in constructing a quaternion to describe a rotation through an angle θ about an axis v', we must choose for φ *half* the angle of rotation: $\varphi = \dfrac{\theta}{2}$, and

$$q = \cos\frac{\theta}{2} + \sin\frac{\theta}{2} \cdot v'.$$

Rodrigues' halving is the reason for spins in physics.

Remark. Because of this halving, a rotation does not determine a quaternion uniquely. The angle θ of rotation is defined modulo $2\pi n$. The angle $\frac{\theta}{2}$ which enters into the definition of the quaternion is defined up to modulo πn. If n is odd, and if we add $2\pi n$ to the angle θ, the quaternion will change sign. We should really write $\pm q = \cos\dfrac{\theta}{2} + \sin\dfrac{\theta}{2} \cdot v'$.

In other words, every rotation corresponds to two quaternions, which differ in sign.

The set of all quaternions of unit norm ($\|q\| = 1$) forms a sphere S^3 in $\mathbb{R}^4$. The correspondence assigning a rotation to each quaternion of unit norm gives a map of the sphere S^3 onto the whole group of rotations which is a double covering: $S^3 \to SO(3) = S^3/\pm 1$.

Theorem. *This map $S^3 \to SO(3)$ is a homomorphism; that is, products are taken into products. If $g(q)$ is a rotation corresponding to the quaternion q, then*

$$g(q_1q_2) = g(q_1)g(q_2). \qquad (*)$$

Thus the multiplication of quaternions is an algebraic description of the multiplication of rotations. (This group is non-commutative.) It is analogous to the theorem that when we multiply complex numbers, we add their arguments. This theorem is proved below.

It is not hard to prove the following statement:

Lemma. $\|q_1q_2\| = \|q_1\| \cdot \|q_2\|$.

(See **[EC2]**.)

It follows that if $\|q_1\| = 1$, then multiplication by this quaternion will conserve length. It is not hard to derive formula $(*)$ from this.

Let z be a quaternion of unit norm: $\|z\| = 1$. We let z act on the quaternion w in the following strange way:

$$w \mapsto zwz^{-1}. \qquad (**)$$

Definition. The *inverse quaternion* of a non-zero quaternion z is the quaternion z^{-1} such that $z^{-1}z = 1$.

This inverse quaternion is unique, and it is not hard to compute it for any nonzero quaternion z in the same way that we compute it for complex numbers (by "inversion"). In our case, $z^{-1} = \bar{z}$, since $z\bar{z} = 1$. (In general, $z^{-1} = \bar{z}/\|z\|^2$.) The strange transformation $(**)$ is a rotation in the space $\mathbb{R}^4$. (The proof is analogous to the case of $\mathbb{C}$; it is given below; see also **[EC3]**.)

If $w = 1$, then this rotation gives $1 \mapsto z \cdot 1 \cdot \overline{z} = z \cdot \overline{z} = \|z\| = 1$; that is, the vector 1 is fixed by transformation $(**)$. If a rotation of Euclidean 4-space $\mathbb{R}^4$ fixes the vector 1, then the orthogonal complement to the vector 1 is also taken onto itself, and our rotation of the space $\mathbb{R}^4$ corresponds to a rotation of the Euclidean space $\mathbb{R}^3$. It can be directly verified that this rotation is the same one that we obtained earlier using the axis and angle of rotation.

Let us examine the action $(**)$ for the product z_1z_2: $z_1z_2w(z_1z_2)^{-1}$. How can we compute $(z_1z_2)^{-1}$?

At this point, I can hear a whisper from the gallery : $(ab)^{-1} = a^{-1}b^{-1}$. False!

Example. Let a represent the action of taking off your jacket, and let b represent the action of taking off your shirt. Then ba consists in taking off your jacket, then taking off your shirt, which is equivalent to getting (partially) undressed. The opposite should of course be getting dressed. You must first put on your shirt, then put on your jacket. Thus, $(ba)^{-1} = a^{-1}b^{-1}$.

Therefore the product z_1z_2 takes the point w into the point

$$(z_1z_2)w(z_1z_2)^{-1} = z_1z_2wz_2^{-1}z_1^{-1} = z_1(z_2wz_2^{-1})z_1^{-1}.$$

Thus the operation $(**)$ of acting by the quaternion z_1z_2 is equivalent to using z_1 to act on the result of z_2 acting on w:

$$(z_1z_2)w(z_1z_2)^{-1} = z_1\left[z_2wz_2^{-1}\right]z_1^{-1}.$$

In other words, for any pure imaginary (orthogonal to 1) quaternion w, we have the identity

$$(g(z_1z_2))(w) = (g(z_1))\left[g(z_2)(w)\right].$$

In a somewhat shorter formulation, we can say that the rotations g, corresponding to quaternions z of unit norm, satisfy the conditions of homomorphism:

$$g(z_1z_2) = g(z_1)g(z_2),$$

so that the difficulties of describing the composition of rotations $g(z)$ reduce to the multiplication of their corresponding quaternions, which is easy to compute.

This method of describing rotations using quaternions is even used in astronautical studies, in arranging the orientation of artificial satellites (sputniks).

We now fulfill our promise to prove that the operation $(**)$ of the action of the quaternion

$$z = \cos\frac{\theta}{2} + v'\sin\frac{\theta}{2}, \quad |v'| = 1,$$

corresponds to a rotation through angle θ about an axis specified by the imaginary part of the quaternion (with unit vector v'), in the space $\mathbb{R}^3$ of pure imaginary quaternions.

(1) The axis with unit vector v' is taken by this operations onto itself. Indeed, the following three obvious facts hold:

(a) Multiplication by the real number $\cos\dfrac{\theta}{2}$ takes any line through the origin onto itself.

(b) Multiplication by the pure imaginary quaternion v' takes the quaternions of the axis with the unit vector v' onto quaternions with an imaginary part equal to zero (since the vector product of any vector with itself is zero).

(c) The inverse quaternion $z^{-1} = \overline{z}$ for the quaternion z has the same form as z (but with the opposite θ), so that multiplication by z^{-1} on the right has properties (a) and (b), as does multiplication by z on the left.

(2) In order to prove that the angle of rotation for (**) about an axis with unit vector v' is equal to θ, it suffices to consider this operation with respect to any purely imaginary quaternion w from the orthogonal complement of the unit vector v' in the three dimensional Euclidean space $\mathbb{R}^3$ of purely imaginary quaternions.

The multiplication of w by z on the left turns this vector within the orthogonal complement mentioned by an angle of $\dfrac{\theta}{2}$ (by the theorem that the argument of the product of two complex numbers is the sum of their arguments). Multiplication by z^{-1} on the right is also a rotation by $\dfrac{\theta}{2}$. (We can deduce this, for example, from the fact that

$$wz^{-1} = w\overline{z} = \overline{z\overline{w}},$$

and under conjugation a purely imaginary quaternion simply changes sign). This proves assertion (2).

Remark. In quantum physics (for example in describing the rotation of electrons), it turns out that that it is not the element of the group SO(3) of rotations that is important, but just one of two quaternions of unit norm which correspond to it, which are known as the *electron spin*, and which has two values (usually denoted in physics as $\pm\dfrac{1}{2}$).

In identifying each pair of opposite points of the sphere S^n as a single point in Euclidean space $\mathbb{R}^{n+1}$ we obtain from the sphere a smooth manifold, called the *real n-dimensional projective space*, and denoted by $\mathbb{R}P^n$ (Figure 29).

This manifold can also be described as the manifold of all lines OM passing through O in the enveloping space $\mathbb{R}^{n+1}$, since such a line is also determined by a pair $\pm N$ of its (opposite) points of intersection with the unit sphere.

Example. The projective line $\mathbb{R}P^1$ is the circle S^1, since if we identify the points φ and $\varphi + \pi$ of the circle $\{\varphi \mod 2\pi\}$, we get a circle $\{\varphi \mod \pi\} = S^1/\pm 1 \approx S^1$.

We can obtain the projective plane $\mathbb{R}P^2$ from the affine plane $\mathbb{R}^2$ by adding to it a "line at infinity" $\mathbb{R}P^1$, containing *one* point at infinity on

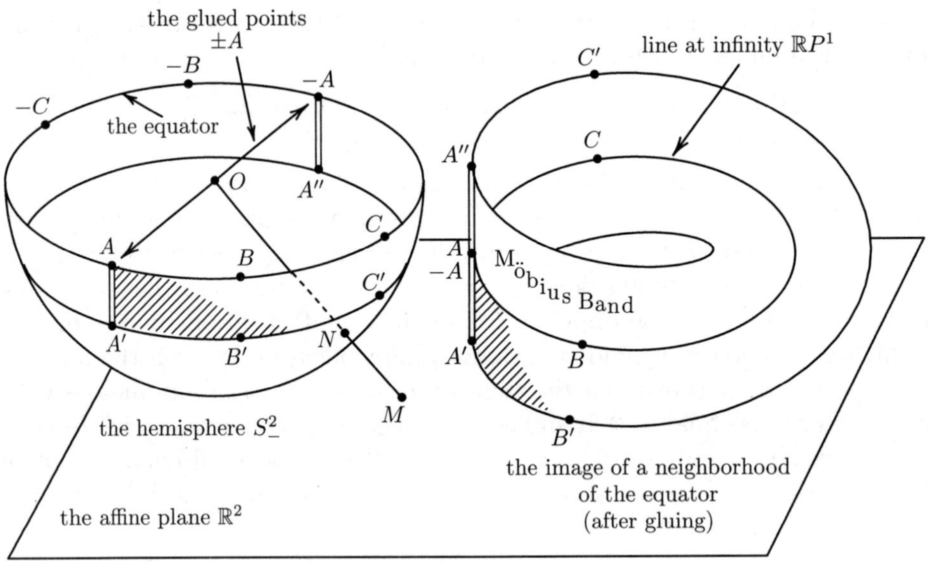

FIGURE 29. Construction of the projective plane by attaching a Möbius band to the disk $A'B'C'$ (or by attaching a projective line $\mathbb{R}P^1$ of points at infinity to the affine plane $\mathbb{R}^2$).

each line of the affine plane. (Moving along a line in either direction, we come to the same point at infinity).

We can see all this clearly if we start by identifying opposite points not on the sphere S^2 but only on the closed hemisphere S^2 (say, the southern hemisphere, below the equator). Then we need only glue together each point A on the equator with its opposite point $-A$, and the "strictly southern" open hemisphere doesn't suffer any changes by gluing (Figure 29).

By the way, it is clear from this same construction that a neighborhood of the line at infinity (and therefore of any line on the projective plane $\mathbb{R}P^2$) is diffeomorphic to a Möbius band (which Möbius himself discovered precisely in this way), as a result of which the projective plane $\mathbb{R}P^2$ is non-orientable (as is every even-dimensional projective space $\mathbb{R}P^{2n}$, unlike the odd-dimensional projective spaces $\mathbb{R}P^{2n+1}$, which are all orientable.)

Thus $SO(3) = S^3/\pm 1 = \mathbb{R}P^3$ is the oriented three-dimensional projective space. We can think of this as of the set of all rotations through all possible angles $0 \leq \theta \leq \pi$, about all possible axes, given by all vectors ω of the unit sphere.

The set of all such rotations can be described as a ball $\{\omega\}$ of radius π in three-dimensional Euclidean space. But on the surface of this ball we must still identify opposite points, since a rotation by an angle π about vector ω coincides with a rotation through an angle π about the vector $-\omega$ (and there are no other pairs of coinciding rotations in our ball).

Exercise. Do there exist automorphisms $A\colon \mathbb{R}^4 \to \mathbb{R}^4$ of the algebra of quaternions; that is, transformations (say, real and linear) for which the relations

$$A(x+y) = A(x) + A(y);\ A(xy) = A(x)A(y)$$

hold for any x and y?

Remark. For the field of complex numbers, the group of automorphisms consists of two elements: $A(z) = z$ and $A(z) = \overline{z}$.

Quaternion conjugation is not an automorphism, since for any quaternions z and w the identity $\overline{zw} = \overline{w} \cdot \overline{z}$ holds, and $\overline{zw}$ does not equal $\overline{z} \cdot \overline{w}$. (Quaternion conjugation is an "anti-automorphism".)

Changing the signs of *two* of the three imaginary components of a quaternion furnishes an example of an automorphism of the algebra of quaternions:

$$A(a + bi + cj + dk) = a + bi - cj - dk.$$

Exercise. Find *all* the automorphisms of the algebra of quaternions (they resemble the one given above.)

Some Examples

Definition. A (real, smooth) transformation of the complex projective plane onto itself is called *pseudoprojective* if it takes every complex projective line into a complex projective line.

Exercise. Prove that *every pseudoprojective transformation of the complex projective plane to itself is either a complex projective transformation, or a product of a complex projective transformation with complex conjugation.* (I do not know whether this is true without assuming smoothness, for example, for homeomorphisms.)

Let us consider a tetrahedron in three-dimensional Euclidean space. The directions from its center to its vertices are given by four points on the real projective plane $\mathbb{R}P^2$. The group of projective transformations which fixes these points coincides with the group A_3 of the symmetries of the tetrahedron, which consists of 24 orthogonal transformations of Euclidean three space $\mathbb{R}^3$ which leave the tetrahedron fixed.

If we embed $\mathbb{R}P^2$ in the complex projective plane $\mathbb{C}P^2$, we obtain four points. To "complexify" the group A_3, let us look at the group of all pseudoprojective transformations of the complex plane which fix this set of four points.

Exercise. Prove that the "complexification of the group A_3 of symmetries of the tetrahedron", as defined above, is the group B_3 of all 48 symmetries of the octahedron, or of the cube, which is the dual of the octahedron in three-dimensional Euclidean space. We can obtain this cube by adding four opposite points to the tetrahedron with its center at the origin.

Remark. We might hope that *the quaternion version of the group* A_3 *of symmetries of the tetrahedron (and also the complex version of the group* B_3 *of symmetries of the cube or the octahedron) would turn out to be the group* H_3 *of 120 symmetries of the icosahedron.* However, this asserts only some strange formulas for the number of edges: these numbers are equal (Figure 30) to $6 = 2 \cdot 3$ for the tetrahedron, $12 = 3 \cdot 4$ for the octahedron (or the cube), $30 = 5 \cdot 6$ for the icosahedron (or dodecahedron)

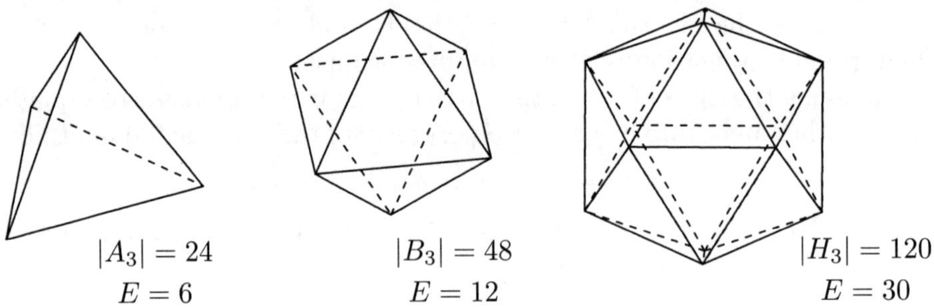

FIGURE 30. The tetrahedron, octahedron, and icosahedron, their symmetry groups, and their number of edges E. The words tetrahedron, octahedron, icosahedron, and dodecahedron, mean "four-faced, eight-faced, 20-faced", and "12-faced".

Each of these three numbers has the form $(n+1)(n+2)$, where $n = 1, 2, 4$ is the the dimension (the manifolds of real numbers, complex numbers, and quaternions, respectively).

The main difficulty in studying the quaternionization consists in the absence of a ready definition (for the case of complex numbers, this role is played by the transition from projective transformations to pseudoprojective transformations.)

We can propose that the role of the pair of tetrahedra inscribed in a cube (and the related fraction $\dfrac{|B_3|}{|A_3|} = \dfrac{48}{24} = 2$) be played, in the case of quaternions, by the five cubes inscribed in a dodecahedron (whose edges are certain diagonals of the pentagonal faces of the dodecahedron–see Figure 31). Under a symmetry of the dodecahedron, these five cubes are permuted, just as the pair of tetrahedra inscribed in a cube are permuted by the action of the group of 48 symmetries of a cube. (And these coincide with the symmetries of the dual octahedron whose vertices are the centers of the faces of the cube.) (See **[EC4]**.)

When we pass from complex numbers to quaternions, the theory of stereographic projections of a sphere onto a plane (with the unavoidable inclusion of Rodrigues' halving) becomes a marvellous parametrization of the projective quaternion line $\mathbb{H}P^1 \approx S^4$ by quaternions, which is analogous to the

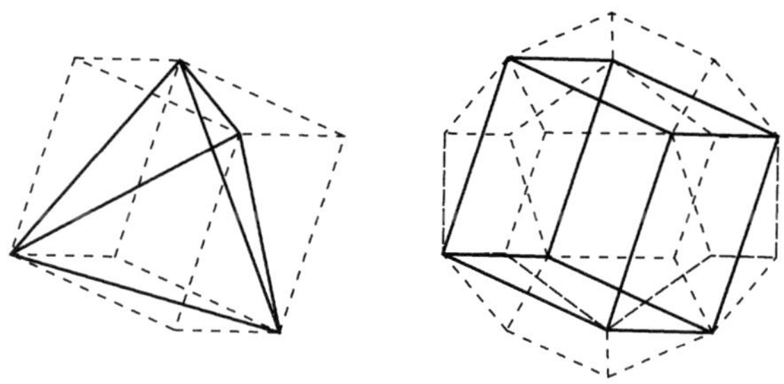

FIGURE 31. A tetrahedron inscribed in a cube, and a cube inscribed in a dodecahedron.

formulas involving "tangents of half angles":

$$\cos\varphi = \frac{1-t^2}{1+t^2}, \sin\varphi = 2t1+t^2, t = \tan\frac{\varphi}{2},$$

which parametrize the circle $\mathbb{R}P^1$ and the Riemann sphere $\mathbb{C}P^1$.

You can read more about quaternion stereographic projection, and about its application to the study of two-sheeted spinor coverings Spin(4) and Spin(5) of the groups SO(4) and SO(5) of rotations of the 3- and 4-dimensional spheres, in the article [**10**].

The formula given above, no matter how strange it may seem, was first discovered during the solution of a famous problem in number theory: finding all "Pythagorean triples" of integers (X, Y, Z) which are the lengths of sides of a right triangle, so that $X^2 + Y^2 = Z^2$.

The simplest cases ($3^2 + 4^2 = 5^2, 12^2 + 5^2 = 13^2$) were used often by the ancient Egyptians, to construct right angles, using a knotted string (for example, in constructing pyramids). But the general formula was made known about one thousand years before Pythagoras, together with the Pythagorean theorem and a proof, on a Babylonian cuneiform tablet of the Chaldeès. Every primitive Pythagorean triple has the form

$$X - u^2 - v^2, Y - 2uv, Z - u^2 + v^2,$$

where (u, v) are relatively prime integers (of opposite parity, so that the triple will be primitive).

On the other hand, these formulas also have topological meaning, as they describe the structure of the set of all complex points of a circle (that is, complex solutions to the equation $x^2 + y^2 = 1$ of the circle–the so-called Riemann surface of the circle). They also give us, as we shall see, conditions for the integrability in elementary functions of the so-called "Abel differentials" of the form $\sqrt{1-x^2}dx$. (General Abel integrals are all integrals of the form $\int R(x,y)dx$ along a curve $H(x,y) - 0$, where H is a polynomial, and R is a rational function.)

Newton's Differential Equation

Newtons' differential equation

$$\frac{d^2x}{dt^2} = F(x),$$

which describes the motion of a point x of unit mass along a line, under the influence of a force field F, has a *first integral of energy*

$$H(x, y) = \text{const}, \text{ where } H = \frac{y^2}{2} + U(x);$$

$y = dx/dt$ is the velocity and U – the potential energy, defined by the condition $F(x) = -dU/dx$.

If F is a polynomial of degree n, then the equation of the law of conservation of energy, $H(x, y) = E$ is defined on the "phase plane" with coordinates x and y of an algebraic (hyperelliptic) curve, which depends on the "constant energy E", while the time t of the motion on the plane is defined by the Abelian integral of the differential form $dt = dx/y$ (since $y = dx/dt$).

The vector field on the phase curve $H = E$ corresponding to this movement of the phase point can be naturally extended to the entire complex Riemann surface (in such a way that $dt = 1$ on the vectors of the field). It is a most surprising fact that this vector field describes the motion along a Riemann surface of the "incompressible fluid" filling it (a fluid having a "zero divergence", which in other terms means that the form dt is closed, i.e. that this form locally is a total differential.)

Question. Suppose the potential energy U is given by a fourth-degree polynomial with two minima ("two potential wells"). Let us investigate the periodic motion in one, then the other of these wells, with identical values for the constant total energy E. The question is, which has the greater period: the motion within the deeper well or the shallower?

Answer. This problem is topological in nature. Both periods are the same. Since the Riemann surface of the phase curve $H = E$ is a torus, the periodic motions are two of its meridians, and the flows of an incompressible fluid on a torus through any two meridians are the same. (See [**EC5**].)

Remark. The integrals of the differential form $dt = dx/y$ along all possible closed paths on a torus, starting at the same point, form a "lattice": the Abelian group $\mathbb{Z}\omega_1 + \mathbb{Z}\omega_2 = \Gamma$, where ω_1 and ω_2 are integrals along the parallel and the meridian of the torus. The value t of the integral along non-closed paths issuing from the given point determine a multi-valued function of the endpoint of the path on the torus, and all the values of this multi-valued function at a point are obtained from any one of them by adding to this value all the numbers from the lattice Γ.

Thus our toroidal Riemann surface can itself be represented (up to a complex diffeomorphism) as a quotient space $\mathbb{C}/\Gamma = \mathbb{C}/(\mathbb{Z}\omega_1 + \mathbb{Z}\omega_2)$, where the complex number $\lambda = \dfrac{\omega_1}{\omega_2}$ is not real.

Taking various fourth-degree polynomials, we can prove that the set of all non-real values λ and all toroidal Riemann surfaces can be obtained by this construction (but the proof of this fact is not easy).

All spherical Riemann surfaces are diffeomorphic, via a complex diffeomorphism, to the standard Riemann sphere $S^2 = \mathbb{C}P^1$.

In light of the tremendous importance of these questions, I will say a few more words about them.

From the Pythagorean Theorem to Riemann Surfaces

Consider the circle $x^2+y^2=1$. We begin by searching for all those point with rational coordinates (x,y). One such point is well-known: $(x=1, y=0)$. Through this point we draw a line with the slope t (that is, the line whose equation is $y=t(x-1)$ (Figure 32).

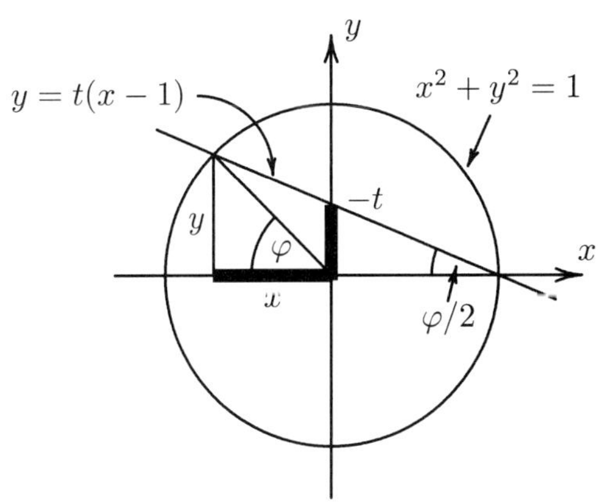

FIGURE 32. Construction of the rational parametrization of the circle by the tangent t of half an angle.

We already know one point of intersection of this line with the circle. So, for a fixed t, we can find the other point, in terms of t, by forming a quadratic equation for the points of intersection of the line and the circle. Therefore the circle is a "rational curve", which admits a parametrization

$$x = P(t),\ y = Q(t), \tag{1}$$

where P and Q are rational functions.

If we make the explicit computation, we quickly find that

$$P = \frac{t^2-1}{t^2+1}, Q = -\frac{2t}{1+t^2}.$$

For rational values of x and y the number $t = \dfrac{y}{x-1}$ is rational, and for rational values of t, we can find rational values of x and y from formula (1). For $-t = u/v$ (with u and v integers) we find the formula for Pythagorean triples given above (where $x = X/Z,\ y = Y/Z$). It is not hard to find fractions that are irreducible (that is, when their numerators and denominators are relatively prime): one just need to avoid the case when u and v are both odd.

On the other hand, the surface formed by *complex* solutions of the equation $x^2 + y^2 = 1$, including "points at infinity", turns out to be, as the parametrization by the complex parameter t shows, the Riemann sphere $S^2 = \mathbb{C}P^1$. (See [**EC6**].)

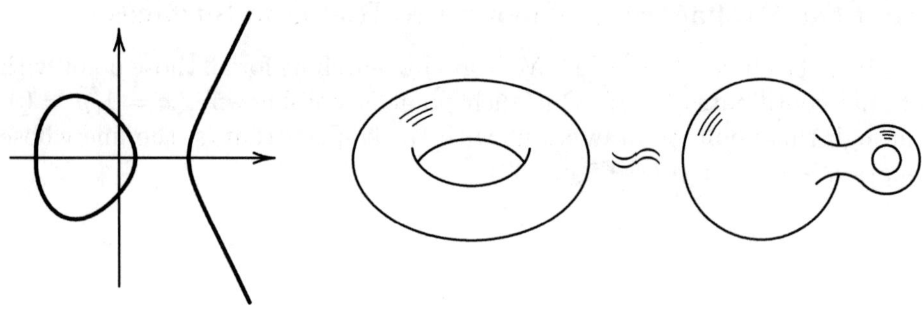

FIGURE 33. Elliptic curve of the third degree: its real points and its Riemann surface.

For other polynomials $H(x, y)$ we would obtain different surfaces for $H = 0$, which might not be spheres. For example, the equation of the "elliptic curve",

$$y^2 = x^3 - x + E,$$

gives the surface of a torus $S^1 \times S^1$ for almost any E. This surface is also called a "sphere with one handle" (Figure 33).

FIGURE 34. Riemann surface of genus $g = 3$: a sphere with three handles.

In more general cases (for example, if we replace the exponent 3 in x^3 with some larger exponent of the form $2g+1$), the Riemann surface will assume the form of a sphere with g handles. (See Figure 34. The number g is called the *genus* of the surface.)

The whole set of closed, connected, smooth, oriented surfaces (without boundaries or singularities) is reduced to spheres with g handles, for $g = 0, 1, 2, \ldots$.

The fundamental theorem of the integration of Abelian differentials along algebraic curves $H(x,y) = 0$ consists in the fact that *such an integral (of any rational form $R(x,y)dx$) can be expressed as an elementary function if and only if the genus of the Riemann surface of the curve $H = 0$ is equal to zero (that is, if it is diffeomorphic to a sphere).*

For example, for $H = x^2 + y^2 - 1$, any such integral, can be expressed by elementary functions, since we have reduced it above to an integral of a rational function in the variable t:

$$\int R(x,y)dx = \int R(P(t), Q(t))P'(t)dt.$$

(This is the "theory of Euler substitutions", the topological nature of which is always hidden from students.)

Any curve of genus 0 is rational; that is, it admits a rational parametrization, analogous to the one we found explicitly for the circle. (It is not too hard to prove the existence of this parametrization, but some familiarity with complex geometry may be needed.)

But if the genus of the Riemann surface of a curve $H = 0$ is greater than zero, then the integral of some rational differential form along this curve is a multi-valued function on the complex plane, with more complicated branching than any elementary function could possibly have. Therefore, in this case the integrals cannot always be expressed in terms of elementary functions.

The simplest example of this kind is the problem of computing the length of an arc of the ellipse, which leads to an integral along a curve of genus $g = 1$ (which is called an *elliptic curve* for this reason).

Thus our theory is connected with complex analysis and topology, with the theory of numbers, and with the theory of the solution of problems in integration using algorithms.

Remark. Exactly the same topological argument occurs in the proof of Abel's theorem about the impossibility of solving the general equation of fifth (or higher) degree in radicals, for example, of the simple equation

$$x^5 + ax + 1 = 0$$

(to which, by the way, any fifth-degree equation can be reduced). In this case, we must look at the branching of a five-valued complex function x with argument a, given by the equation above, as the complex variable a goes around a branching point of this five-valued function.

The branching of this function can be described by permutations of five local sheets $x_j(a)$. Those permutations, which correspond to all possible paths which don't pass through the branching point, form a group. We call this group the "monodromy group" of the multi-valued function x of the argument a (Figure 35).

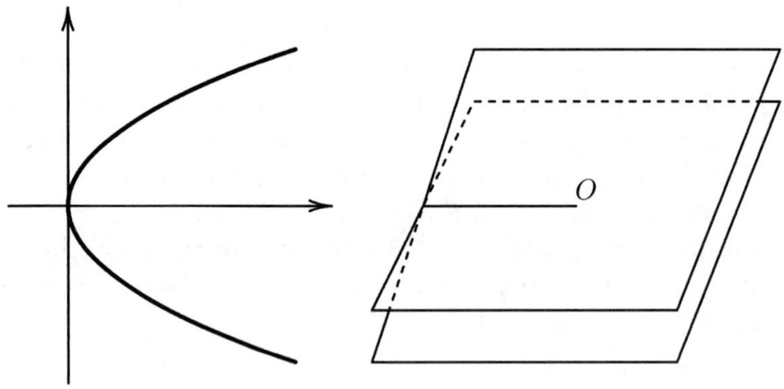

FIGURE 35. The monodromy group of the "square root" function consists of two permutations of the two roots.

The monodromy group of any combination of radicals is solvable (it is reduced to the skew product of commutative groups). But the monodromy group of the five-valued function considered above contains all permutations of five elements, and is therefore not solvable. This means that our quintic equation is also not solvable in radicals (even if we augment the set of radicals with the set of all single-valued functions). You can read about this theorem of Abel in more detail in the book by V. B. Alekseev [**1**], which is an account of my lectures to high school students in 1964.

The genus g of a curve is not all that easy to find, even if its equation $H(x, y) = 0$ is given. For this, an invention of "Italian algebraic geometry" is helpful: if we fix the degree n of a polynomial h, then *its genus will be the same for almost any values of the coefficients of the polynomial.* Those special polynomials whose genus is different are quite rare. Just as there are values of the parameter c for which the equation $x^n = c$ has an atypical number of roots (a number not equal to n), these special polynomials form a set given by a complex equation ("the discriminant is zero"); that is, they are given by two real equations. (In the example given above, this equation is $-c = a + bi = 0$, which leads to the system $(a = 0, b = 0)$.)

Therefore, *the "non-special" polynomials H form a connected set in the space of all polynomials of given degree n* such that as we move along it, any topological invariant (for example, the genus) corresponding to the moving point on the Riemann surface remains unchanged.

It remains to determine the genus g for a *single* non-special example. This is not so difficult to do: it is enough, for example, to take the nearly-degenerate curve $H = 0$, where $H = L_1 \cdot \ldots \cdot L_n + \varepsilon$, and where the L_j

are generic non-homogeneous linear functions. This Riemann surface can be obtained from n spheres by attaching N tubes, where N is the number of intersections of n lines (Figure 36; see also [**EC7**]).

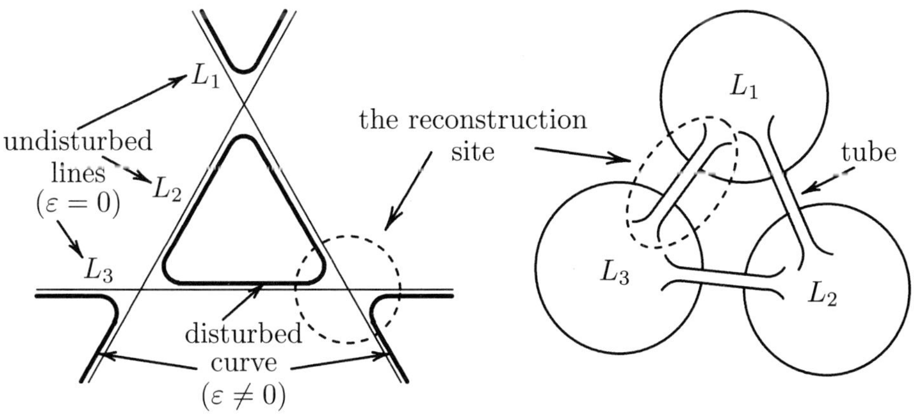

FIGURE 36. The topology of a real algebraic curve and the Riemann surface of the nearly-degenerate algebraic curve of degree 3 (before the construction, the number of spheres here is $n = 3$, the number of connecting tubes is $N = 3$, and the genus $g = 1$).

The answer is called the *Riemann-Hurwitz formula*:

$$g = \frac{(n-1)(n-2)}{2}.$$

To remember it, notice that the genus of curves of degree $n = 1$ (which are lines) and $n = 2$ (ellipses, hyperbolas, parabolas) is equal to zero; that is, these curves admit of a rational parametrization, and their Riemann surfaces are diffeomorphic to a sphere.

Rational curves are also called *unicursal* (that is, "singly described"), since in the real case, one can draw them with one stroke of the pen, without lifting it from the paper. (In this situation we must consider the curve as drawn on the *projective* plane, or else we must assume that the real part of the curve, in an affine plane, is *compact*: the ellipse is unicursal, but the affine part of a hyperbola is not).

The proof of the fact that a rational curve is unicursal can be obtained simply from an examination of the real values of a rationally parametrized curve with variable t.

If the initial curve $H = 0$ is not smooth, then each singular point decreases the genus of the corresponding Riemann surface (on which, for example, the point where the curve has a simple self-intersection corresponds to two points "on different sheets"). It turns out that a curve of degree n with $\dfrac{(n-1)(n-2)}{2}$ singular points (and there cannot be more than these) already has genus 0. That is, it is rational and unicursal, and the integrals of rational forms along it can be computed using elementary functions. An example is the "degenerate elliptic curve" (Figure 37) given by $y^2 = x^3 - 3x + 2$ (with a singular point, a simple self-intersection, at $x = 1, y = 0$).

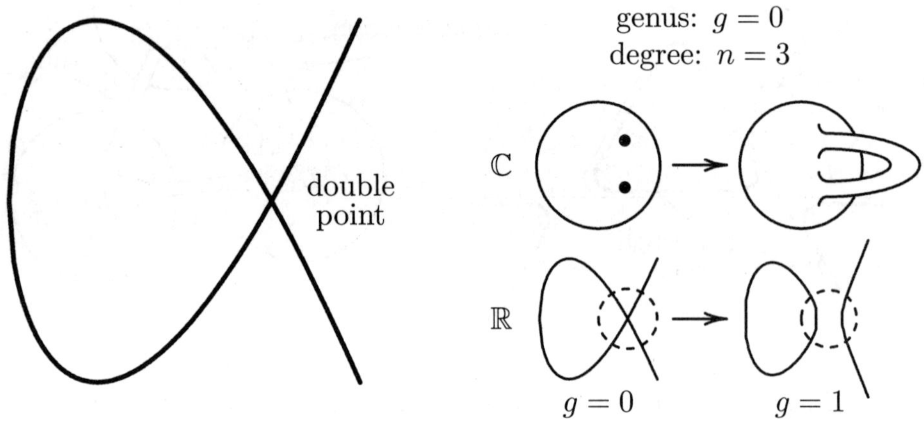

FIGURE 37. A singular unicursal curve, along which any Abelian integral can be taken, and its minor reconstruction as a non-special elliptic curve (in the real or complex domain).

Remark. Instead of using the parameter t to form rational expressions to represent the coordinates x and y of our unicursal curve, we can use the slope of the line connecting the point we are discussing with the point of self-intersection. (We can compute $x(t)$ and $y(t)$ explicitly. Then this will permit us to compute any Abelian integral along our curve.)

By the way, this construction explains why a self-intersection point on the Riemann surface of a curve turns into two ordinary points. The two values, t_1 and t_2, of the parameter t, corresponding to the point of self-intersection of the original curve are the slopes of the two smooth branches of the original curve at the point of self-intersection.

The transition from a curve with self-intersections to its smooth Riemann surface, on which a point of self-intersection is represented several times, is called a *normalization* of the curve.

It turns out that any algebraic curve can be algebraically normalized. That is, any algebraic curve can be obtained from some smooth algebraic curve (on its Riemann surface) by an algebraic transformation (which, however, might sent certain distinct points on the Riemann surface onto some singular points of the image curve).

Exercise. Normalize the lemniscate $y^2 = x^2 - x^4$. Is this curve unicursal?

Mathematical Trinities

Many mathematical theories have three versions: real, complex, and quaternion. Sometimes it is not easy to recognize the unity either of the corresponding theorems, or of their applications (be it to topology, to physics, to number theory, or to algebra).

I will give just a few examples.[11]

Example 1. The coincidence of the projective plane and the circle:

$$\mathbb{R}P^1 = S^1.$$

The complexification of this fact turns out to be a wonderful theorem of Pontryagin (discovered by him in the 1930's, but not published, and thus now known in the West by the names of those mathematicians who published their proofs in the 1960s, as an answer to my question as to whether they knew a proof of this theorem of Pontryagin's).

Theorem. *The quotient space of the complex projective plane over its real diffeomorphism of "complex conjugation" is diffeomorphic to the four-dimensional sphere*:

$$\mathbb{C}P^2/\mathrm{Conj} \approx S^4.$$

Thus, in the complexification, the dimension (one) of the projective space becomes two, and in addition we must factor by the group of automorphisms of the field of complex numbers (which we could have done in the real case, where, however, the only automorphism is the identity transformation).

It is difficult to guess at the quaternion analogue of the preceding theorem, but an analysis of the logic of its proof reveals the following:[12]

$$\mathbb{H}P^4/\mathrm{Aut}/\mathrm{Conj} \approx S^{13}.$$

Here we must start with the projective space of quaternion dimension four (thus, of real dimension sixteen), and factorize by the three-dimensional group of automorphisms (isomorphic to $SO(3)$) and also by the antiautomorphism of quaternion conjugation.

It is instructive, however, that the proofs of the three facts listed above are parallel. It is enough to replace real numbers with complex numbers (and quaternions), and replace the quadratic forms (which in the real case can be written, using an appropriate coordinate system, in the form $\Sigma_{m=1}^{n} a_m x_m^2$) with real Hermitian (or, hyper-Hermitian) forms, which can be written in the same way, just by changing the squares x_m^2 into the squares of moduli $|x_m^2|$.

By definition, Hermitian and hyper-Hermitian forms (in the complex vector space $\mathbb{C}^n$ and quaternion space $\mathbb{H}^n$) are ordinary real quadratic forms

---

[11] A more detailed discussion of a larger number of these facts can be found in the article [**8**].

[12] See the article [**9**].

(correspondingly, in $\mathbb{R}^{2n}$ and $\mathbb{R}^{4n}$), which are invariant with respect to multiplication of the vector argument by complex numbers (quaternions) of unit norm.

The geometric object corresponding to a positive definite quadratic form f is the ellipsoid $f = 1$. Thus the Hermitian (in the quaternion case, hyper-Hermitian) forms correspond to ellipsoids of revolution with special symmetries: they are taken into themselves by multiplication of all vectors of the space of the ellipsoid by i in the complex case (and by i, j, k in the quaternion case).

Now I can describe a second example of a wonderful quaternionization.

Example 2. The repulsion of electronic levels, Hall's quantum effect, and characteristic numbers.

Even in the real case, the result is not at all trivial and was discovered, despite its fundamentally mathematical nature, only as a result of the development of quantum mechanics (where it is called the theory of von Neumann-Wigner). Consider the manifold of all ellipses (with their centers at the origin) in the Euclidean plane (or, if it helps, the manifold of quadratic forms which determine them).

Some of the ellipses are actually circles. At first glance it may seem that the condition that an ellipse be a circle is the single condition of equality of the two semiaxes, $a = b$, so the submanifold of circles must have codimension one in the manifold of all ellipses.

But this is not the case: the manifold of quadratic forms

$$Ax^2 + 2Bxy + Cy^2$$

has dimension 3 (and coordinates A, B, C), while the set of circles has dimension 1 (since a circle with the center at the origin is defined simply by its radius).

The condition "the discriminant is zero" (for the quadratic equation defining the lengths of the semiaxes of the ellipses), which singles out circles, has the form $(A + C)^2 = 4(AC - B^2)$; that is, it reduces to the sum of two squares: $(A - C)^2 + 4B^2 = 0$, and defines a one-dimensional subset of the three-dimensional space of forms (namely, the line $A = C, B = 0$).

The theorem of von Neumann-Wigner asserts that even for ellipsoids in n-dimensional space, for any n, *the submanifold of ellipsoids of revolution has codimension* 2. In other words, *not only an ellipsoid in general position is not an ellipsoid of revolution, but also a one-parameter family of ellipsoids in general position does not contain any ellipsoid of revolution.*

If we draw the graph of the dependence on the parameter p of the lengths of the n semiaxes $a_m(p)$ for an ellipsoid of such a family, , then we will get n curves ($m = 1, \ldots, n$) on the (p, a)-plane , each of which has a one-to-one projection onto the axis of values of the parameter p, and which are all disjoint, although they may sometimes creep close to one another (Figure 18).

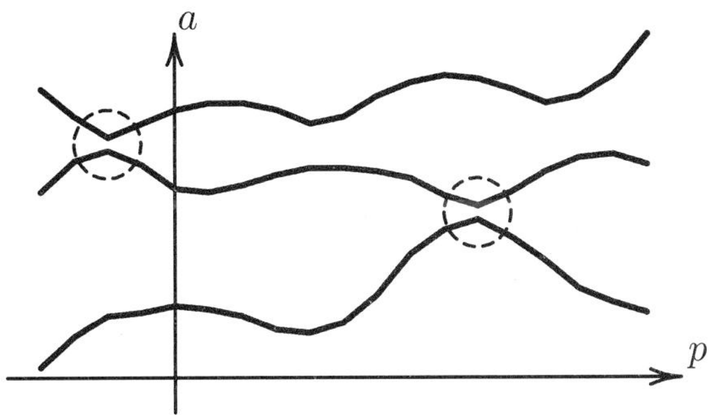

FIGURE 38. Repulsion of the eigenvalues.

In physics, the values a_n are called "levels", and the fact that they remain distinct can be interpreted as a "repulsion" of the levels from one another as they approach each other with the changing parameter.

By the way, since this theorem is a mathematical statement, it has many of physical (and other) applications. For example, as a satellite rotates about its center of mass its "ellipsoid of inertia" exerts a strong influence. If that ellipsoid turns out to be an ellipsoid of revolution, then control of the orientation and tumbling of such a satellite is easier. The theorem of Wigner-von Neumann shows that *in order to make the ellipsoid of inertia of a satellite into an ellipsoid of revolution, it is not sufficient to move just one "calibrating weight" along a crossbar: at least two crossbars are necessary.*

We now turn to the complexification of the theorem about repulsion of levels. When we go from the quadratic forms of $\mathbb{R}^n$ to the Hermitian forms of $\mathbb{C}^n$, the codimension two of the manifold of ellipsoids of revolution within the manifold of all ellipsoids is changed into a real codimension of three.

Ellipsoids of revolution are absent not just in single-parameter families in general position, but also in two-parameter families. (And in families of three real parameters, we can find Hermitian ellipsoids with additional symmetry for certain points in the three-dimensional space of parameters.)

I have investigated the topological questions arising here in detail (the structure of the vector bundle of "eigenvectors", which correspond in mathematics to the major axes of an ellipsoid, and are called "modes" in physics) in the article [**3**].

Today these results are called the theory of the "integer quantum Hall effect" (because there is the possibility of experimental observation of the passage of the surface in three dimensional space-time, whose point is determined by two parameters, through the special points, where the corresponding ellipsoids have an additional symmetry).

In topological terms this phenomenon corresponds to a modification of the Chern characteristic number of the complex vector bundle of eigenvectors over the surface when this surfaces passes through the special points. Thus, the topological theory of the integer quantum Hall effect was constructed and published in 1972, before the appearance of the physical theory, simply by the complexification of the classical theorem of Wigner and von Neumann. The only thing missing was the terminology of physics, which was supplied later on.

Today we are in the same situation with regard to the quaternion version of the theory of Wigner and von Neumann as we were in 1972 with the complex version.

In this case, the real codimension of the phenomenon of additional symmetries (that is, the collision of the "eigenlevels" of hyper-Hermitian forms) is equal to five. We have already encountered these numbers 2, 3, 5 when we analyzed the numbers of edges in a tetrahedron, octahedron, and icosahedron: each is equal to $d+1$, where d is the real dimension (of a real, complex, and quaternionic line).

The role of the Hopf fibration $S^3 \to S^2$ (with fiber S^1), which is fundamental to the complex theory of quantum Hall effect, is played in the quaternion case, by the "second Hopf fibration" $S^7 \to S^4$ (with fiber S^3).

Both these fibrations can be described using the standard construction of projective space ($\mathbb{C}P^1 \approx S^2, \mathbb{H}P^1 \approx S^4$), from the sphere of the corresponding projective vector space ($S^3 \subset \mathbb{C}^2 \backslash 0, S^7 \subset \mathbb{H}^2 \backslash 0$). These fibrations assign to a point of the sphere the line connecting this point with zero. And finally, the "Chern characteristic classes and number" in the complex case (which are the complexification of the "Stiefel-Whitney characteristic classes" and the "Euler characteristic" in the real case) correspond in the quarternion case to the "Pontryagin characteristic classes and numbers" (which are subject to a modification when the moving four-dimensional space of parameters passes through special points in five-dimensional space).

The only problem is that we have not yet arrived at physical nomenclature for these ready-made mathematical results (although I hope that the abundance of various fields and particles in contemporary physics will allow us to specify those experimentally observable situations, which are governed by the quaternion theory of hyper-Hermitian matrices described above). The trios $\mathbb{R}$, $\mathbb{C}$, $\mathbb{H}$ and "Stiefel-Whitney classes, Chern classes, Pontryagin classes" specify the complexification of the two-element group $\mathbb{Z}_2$ of the coefficients of the Chern classes. However, what the quaternionic version of the group $\mathbb{Z}_2$ (or the complex version of the group $\mathbb{Z}$ of integers) will turn out to be has not yet been decided. (Possible candidates are, in particular, the groups $\mathbb{Z}$ of integers and the group $\mathbb{Z}+i\mathbb{Z}$ of complex integers.)

Spins and Braids

For the description of the double covering of the group of rotations $SO(3)$ by the group of spins $Spin(3) = S^3 = SU(2)$, the physicists have thought up a beautiful method, based on the mathematical *theory of braids.* Let a set of n pairwise distinct points be fixed on a surface n. A braid of n strands on a surface M is defined as a path in the space of such sets, starting and ending in the given (unordered) set of n points.

More precisely, a braid is such a path defined up to a homotopy, that is, up to a continuous deformation preserving the initial and the terminal sets and, of course deforming the intermediate sets in such a way that at every moment they consist of n different points. In other words, we consider that such a deformation does not change a braid.

We can construct a group of braids of n strands, by continuing one path by the other one. (The inverse braid is a path, traversed it the opposite direction).

For example, there is a braid a of two strands on an oriented plane which is a motion of these points by a half-turn in the positive direction about the center of the segment joining their midpoints. All braids a^k, $(k = \pm 1, \pm 2, \dots)$ are distinct, non-trivial (different from the motionless path $a^0 = e$), and together with e form a group $B(2)$ of all braids of two strands on the plane (which is isomorphic to the group of integers $\mathbb{Z}$.)

The group $B(3)$ of all braids of three strands on the plane is generated by two transformations a and b, which rotate, respectively, the pair of points I, II and the pair of points II, III from the set of points I, II, III lying (in this order) on a line (Figure 39). These two generators satisfy the relation $aba = bab$ (the reader can see this in Figure 39), but no other relations (the proof of this is more difficult).

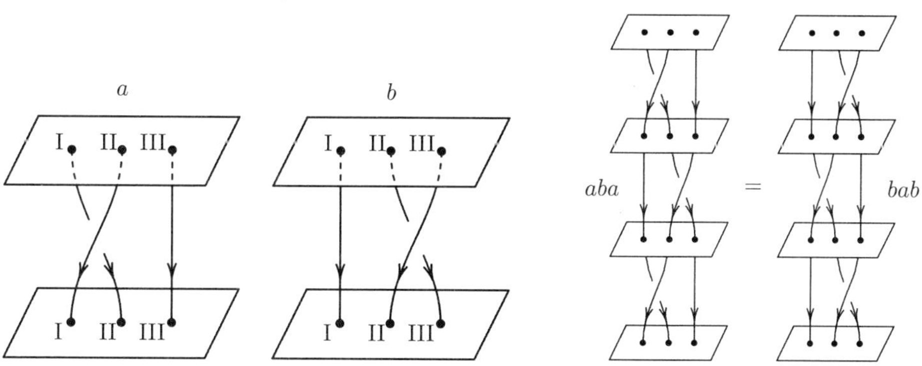

FIGURE 39. Three-strand braids on the plane.

It is convenient to represent braids on a plane by curves in three-dimensional space-time: a braid of n strands consists of n such curves (which are called strands). By tradition, the x-axis is drawn vertically and oriented from above to below. (This tradition can be explained by the usual way of weaving a braid, from above to below.)

The idea of explaining the double covering of rotations by spins consists in looking at braids of n strands on the sphere S^2 (for example, for $n = 4$).

In the group of braids on the plane there are no elements of finite order: if the mth power of some braid is trivial, then the braid itself is trivial. For a braid of two strands this is obvious, but even for three strands the proof is not at all simple.

It is an amazing mathematical fact of the theory of spherical braids that the group of braids on the sphere S^2 has elements of finite order (for example, of the second order, in the case of braids of four strands).

This topological result is very close to the fact that the fundamental group of the variety $SO(3) \approx \mathbb{R}P^3$ consists of two elements, so that the spinor covering $S^3 \to SO(3)$, which assigns to a quaternion of unit norm the rotation determined by this quaternion, is of two sheets.

P. Dirac, who invented this method of explaining spins, demonstrated to the physicists an experimental proof of the theorem we've formulated about spherical braids. To do this, he made a model of the appropriate spherical braid of four strands, tying two concentric spheres by four ropes in the layer bounded by them. (This layer replaces, in the spherical situation, the three-dimensional space-time, which contains the strands of a braid corresponding to the motion of a set of n points on the plane.)

Then, inside the ball bounded by the smaller sphere, we place a still smaller sphere, connected to the original smaller (but now middle-sized) sphere in the same way as it was connected with the largest sphere.

And finally, we get rid of the middle sphere. After that, the largest and smallest sphere end up connected by 4 ropes in a trivial way (after a deformation, the ropes become radial), although the original connection was not at all trivial and could not be untangled (cannot be made radial by a deformation in the layer between the two concentric spheres that bound it.)

Unlike physicists, mathematicians usually do not know this theorem from the theory of spherical braids, since they are not interested in spins.

Appendix

Definition. The function, which takes a non-zero complex number z into

$$F(z) = \frac{1}{2}\left(z + \frac{1}{z}\right),$$

is called the *Zhukovksy function.*

Theorem 1. *The Zhukovsky function transforms the circle* $|z| = r$ *of the complex variable* z *into an ellipse centered at 0 and with semiaxes* a, b, *where* $2a = r + r^{-1}, 2b = r - r^{-1}$, *on the plane of the complex variable* $w = F(z)$.

Proof. The real and imaginary parts of the number z are equal to $r\cos\phi$ and $r\sin\phi$, so

$$\frac{1}{z} = \frac{1}{r}(\cos\phi - i\sin\phi),$$

whence

$$w = \frac{1}{2}\left[(r + r^{-1})\cos\phi + i(r - r^{-1})\sin\phi\right],$$

which is what we had asserted.

The Zhukovsky function plays an enormous role in engineering, and Zhukovsky introduced it in investigating the lift force of an airplane's wing (where the formula of Zhukovsky, based on it, remains fundamental).

But it also has many other applications.

Theorem 2 (Bohlin). *The transformation of squaring complex numbers, $\mathbb{C} \to \mathbb{C}$, which takes z onto $w = z^2$, takes a Hooke ellipse with its center at 0 into a Newtonian ellipse with a focus at 0.*

Proof. Suppose the complex number u moves around the circle $|u| = r$. Then the complex number $z = u+u^{-1}$, by Theorem 1, runs through a Hooke ellipse with semiaxes $2a$ and $2b$ (and with center at 0). Thus we can obtain an ellipse with any ratio of its semiaxes. The square of the number z is

$$z^2 = u^2 + u^{-2} + 2. \tag{1}$$

The first two terms in the sum again describe a Hooke ellipse (also by Theorem 1). Its semiaxes have lengths $2A = r^2 + r^{-2}$ and $2B = r^2 - r^{-2}$. The square of the distance from the center of this ellipse to its focus can be computed by the Pythagorean Theorem:

$$(2A)^2 - (2B)^2 = [(r^2 + r^{-2})^2 - (r^2 - r^{-2})^2] = 4.$$

That is, the distance from the center of the ellipse $\{u^2 + u^{-2}\}$ to its focus is $2C = 2$.

Therefore, after we make a shift by 2 in accordance with formula (1), the origin will become a focus of the shifted ellipse. This proves the theorem.

Remark 1. We can reformulate this result in physical terms: the trajectories of harmonic oscillation about the origin on the complex plane provided by Hooke's law are transformed by the squaring transformation into the trajectories of motion provided by the law of universal gravitation (or by Coulomb's law of attraction) whose force is inversely proportional to the square of the distance from the center of attraction. The motions themselves are not transformed into each other: the velocities of passage around an orbit are different for the two motions.

Remark 2. The theorem of Bohlin proven above has a remarkable generalization to the case when squaring is replaced by raising to some other power α. In this case, the orbits of motion in the field of attraction (or repulsion) of degree A are transformed into the orbits for a field of degree B, where the numbers A and B are related by a principle of duality:

$$(A + 3)(B + 3) = 4.$$

For example, Hooke's law, for which $A = 1$, is the dual to the law of universal gravitation, for which $B = -2$. In place of the exponent α in the general case we must choose the value $\alpha = (A+3)/2$. (The dual law corresponds to the inverse transformation, with the inverse dimension $\beta = (B+3)/2 = 1/\alpha$.

Problem. Find all self-dual laws. (Newton has already investigated these!)

In a remarkable way, this whole theory of duality extends to the case when the power transformation $w = z^\alpha$ is replaced by an arbitrary polynomial (or even by an arbitrary smooth complex function in a complex domain $w(z)$, such as $w = e^z$ or $w = \ln z$).

The dual potentials on the planes $\{z\}$ and $\{w\}$ are given by formulas of two real functions of a complex variable, determined by the transformation:

$$U(z) = |dw/dz|^2, V(w) = -|dz/dw|^2.$$

This transformation takes the orbits of motion of point z with constant total energy E in the field with potential energy U into the orbits of motion of point w with constant total energy $-1/E$ in the field with potential energy V.

This amazing duality of motions in such different two-dimensional fields also holds for Schrödinger's equation in quantum mechanics.[13] But neither generalization of this duality to the motion in spaces of other dimensions, nor quaternionic generalizations of this duality are known.

[13]See [**21**].

Editors' Comments

[EC1] Consider vectors $\overrightarrow{OA_1}$, $\overrightarrow{OA_2}$,$\overrightarrow{OB}$ and $\overrightarrow{OA} = \overrightarrow{OA_1}+\overrightarrow{OA_2}$ (see Figure 40). We want to prove that the area of parallelogram $OACB$ is equal to the sum of the areas of parallelograms OA_1C_1B and OA_2C_2B.

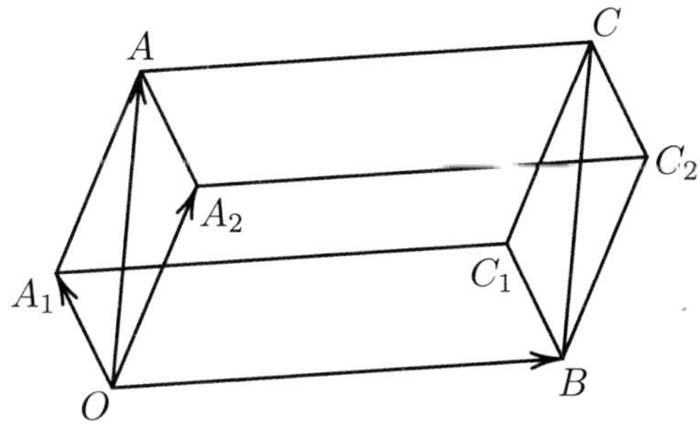

FIGURE 40.

Our drawing contains two congruent triangles, OA_1A and BC_1C, and two congruent parallelograms OA_2C_2B and A_1ACC_1. Using these, we obtain:

$$\begin{aligned}\text{area}(OACB) &= \text{area}(OACB) + \text{area}(OA_1A) - \text{area}(BC_1C)\\ &= \text{area}(OA_1ACC_1B)\\ &= \text{area}(OA_1C_1B) + \text{area}(A_1ACC_1)\\ &= \text{area}(OA_1C_1B) + \text{area}(OA_2C_2B).\end{aligned}$$

Notice that the drawing and the computations will be a bit different, if the direction of the vector $\overrightarrow{OB}$ is *between* the directions of vectors $\overrightarrow{OA_1}$ and $\overrightarrow{OA_2}$. In this case, the (signed) areas of the parallelograms OA_1C_1B and OA_2C_2B will have opposite signs, and the equality to prove will be area(OA_1C_1B) − area$(OA_2C_2B) = \pm$ area$(OACB)$. The proof will be basically the same.

[EC2] Proof of Lemma:

$$\begin{aligned}\|q_1q_2\|^2 &= q_1q_2\overline{q_1q_2} = q_1q_2\overline{q}_2\overline{q}_1 = q_1(q_2\overline{q}_2)\overline{q}_1 = q_1\|q_2\|^2\overline{q}_1 = \|q_2\|^2q_1\overline{q}_1\\ &= \|q_2\|^2\|q_1\|^2 = (\|q_1\|\cdot\|q_2\|)^2.\end{aligned}$$

[EC3] We want to prove that if $\|z\| = 1$, then for any purely imaginary quaternion w, $[g(z)](w) = zwz^{-1} = zw\overline{z}$.

First, notice that the transformation $w \mapsto zw\overline{z}$ preserves norms: $\|zw\overline{z}\| = \|z\|\,\|w\|\,\|\overline{z}\| = \|w\|$. Hence, this transformation also preserves angles. In particular, it takes orthogonal vectors into orthogonal vectors.

Furthermore, it takes v' into v'. Indeed, $(v')^2 = -v'\overline{v}' = -\|v'\| = -1$ and hence

$$\begin{aligned}zv'\overline{z} &= (\cos\varphi + \sin\varphi\, v')v'(\cos\varphi - \sin\varphi\, v')\\ &= \cos^2\varphi\, v' + \cos\varphi\sin\varphi(v')^2 - \cos\varphi\sin\varphi(v')^2 - \sin^2\varphi(v')^3\\ &= \cos^2\varphi^2\, v' + \sin^2\varphi\, v' = v'.\end{aligned}$$

Also, it takes purely imaginary quaternions into purely imaginary quaternions: it obviously takes 1 into -1, and hence takes any quaternion orthogonal orthogonal to 1 into a quaternion orthogonal to -1. (Or, we can see that if $\overline{w} = -w$, then $\overline{zw\overline{z}} = \overline{\overline{z}}\,\overline{w}\,\overline{z} = -zw\overline{z}$.)

Thus, the transformation $w \mapsto zw\overline{z}$ determines an orthogonal transformation of the space of purely imaginary quaternions which fixes a vector v' of this space. Hence, it is a rotation of this space about the line of vector v', and it remains to find the angle of this rotation.

In the space of pure imaginary quaternions, consider the (two-dimensional) orthogonal complement P of v'. Vectors from P are orthogonal to v' and to 1, hence, they are also orthogonal to z.

Take a unit vector $w \in P$. Then zw also belongs to P. Indeed, $\operatorname{Re}(zw) = \operatorname{Re} z \cdot \operatorname{Re} w - (z, w) = 0$, so zw is purely imaginary; on the other hand, $(zw, v') = \cos\varphi(w, v') + \sin\varphi(v'w, v') = \sin\varphi(v'w, v')$ and $(v'w, v') = (v', v'w) = \operatorname{Re}(v')\operatorname{Re}(v'w) - \operatorname{Re}(v'v'w) = -\operatorname{Re}(v'v'w) = \operatorname{Re} w = 0$. In the same way, we can see that $wz, \overline{z}w$ and $w\overline{z}$ belong to P as well.

The transformation $w \mapsto zw$ is a rotation of the plane P, but through what angle? Assuming that $\|w\| = 1$, we find that $(zw, w) = \operatorname{Re}(zw)\operatorname{Re} w - \operatorname{Re}(zw^2) = -\operatorname{Re}(zw^2) = -\operatorname{Re}(-z) = \cos\varphi$. Thus, $w \mapsto zw$ is a rotation of P in the positive direction by the angle φ. In the same way, we see that the transformation $w \mapsto w\overline{z}$ is also a rotation of P in positive direction by the angle φ. Thus, the transformation $w \mapsto zw\overline{z}$ is a product of two such rotations, so it is a rotation by the angle 2φ, which is θ, if $\varphi = \dfrac{\theta}{2}$.

[EC4] To be more precise, there are 120 symmetries of the dodecahedron and 5 cubes inscribed in the dodecahedron. Therefore, every inscribed cube is mapped into itself by $\dfrac{120}{5} = 24$ symmetries of the dodecahedron. However, the full group B_3 of symmetries of the cube has 48 elements, which shows that the restrictions of the symmetries of the dodecahedron compose a subgroup of index 2 of B_3. Actually, it is not hard to prove that this subgroup consists of two kinds of symmetries of the cube: those which preserve each of the inscribed tetrahedra and preserve the orientation and those which switch the inscribed tetrahedra and reverse their orientation.

[EC5] This paragraph contains references to several deep geometric facts, which the reader may wish to think about. First, for every point (x, y) of the "phase space" (that is, of the real plane P), x is the initial position of the particle on the line, and y is the initial speed of this particle. (It may be easier to think of this line as a frictionless slide in the shape of the graph of the function $y = F(x)$.) Then x and y determine a motion $(x(t), y(t) = x'(t))$; that is, a parametrized curve in P (with $x(0) = x, y(0) = y$). The law of conservation of energy then says that the total energy H is constant on this curve. (For our example, all such curves $H = E$ are closed). Next, we complexify and projectivize the whole construction. This makes the curves $H = E$ into surfaces in the complex projective plane $\mathbb{C}P^2$, and our vector field becomes two vector fields, one on each of these surfaces. Moreover, the surfaces are actually, tori, and the vector fields are divergence-free. The time which our particle needs to make a full rotation about the closed curve $H = E$ becomes the flow of the vector field through a closed curve on the torus. It follows from Stokes' Theorem that for a divergence-free vector field, the flow through any closed curve bounding a domain is zero. So the flows through two curves which both bound a domain are the same. In particular, the flows through two meridians of the torus (as well as through two parallels) are the same. Also, the flow through any closed curve on the torus has the form $k_1\omega_1 + k_2\omega_2$ where k_1 and k_2 are integers, and ω_1 and ω_2 are flows through a meridian and through a parallel.

[EC6] Geometrically, the parametrization of the "complex circle" $x^2+y^2 = 1$ has the same meaning as the parametrization of the real circle shown in Figure 32. We take the point $(x, y) \in \mathbb{C}^2$, draw a (complex) straight line through (x, y) and $(1, 0)$, then find the intersection point of this line with the y axis. This point is $(-t, 0)$, where t is the parameter value corresponding to (x, y). In geometry, this construction (with t instead of $-t$) is called the *stereographic projection.* The point $(1, 0)$ itself does not correspond to any value of t, or rather corresponds to the infinite value of t. In the complex case, the point $(0, 1)$ also corresponds to the infinite value of t. However, the complex circle, unlike the real circle, has two points at infinity itself, which correspond to the parametric values $t = \pm i$.

[EC7] In $\mathbb{C}^2$, the "curve" $L_1 \cdots L_n = 0$ is the union of n complex lines (which look like 2-dimensional planes in $\mathbb{R}^4$ from the point of view of real geometry). In the projective space $\mathbb{C}P^2$, each plane acquires a point at infinity, so the planes become 2-dimensional spheres, every pair of which crosses each other at one point. If we perturb the curve to become $L_1 \cdots L_n = \varepsilon$ (with a small $\varepsilon \neq 0$), then the curve will not suffer any significant changes in the complement of a small neighborhood of the intersection points $\{L_i = 0\} \cap \{L_j = 0\}$ $(i \neq j)$. Let us investigate what happens in these neighborhoods.

Consider the point $P = \{L_i = 0\} \cap \{L_j = 0\}$ Without loss of generality, we may assume that $i = 1$, $j = 2$, and $L_1(x, y) = x$, $L_2(x, y) = y$ (so $P = (0, 0)$). In a small region near P, the curve $L_1 \cdots L_n = \varepsilon$ is very close to $xy = \varepsilon'$ where $\varepsilon' = \varepsilon/(L_3(0, 0) \cdots L_n(0, 0))$. Let $\tau = \sqrt{\varepsilon'}$ (we can choose any value of the square root). An arbitrary point (x, y) of our curve has the form $(\lambda\tau, \mu\tau)$ with $\lambda\mu = 1$. The curve falls into two parts: $|\lambda| \geq 1$ and $|\mu| \geq 1$, which share a circle $|\lambda| = |\mu| = 1$. (This circle consists of points $(\lambda\tau, \bar{\lambda}\tau)$ with $|\lambda| = 1$.) The domain $|\lambda| \geq 1$ is, topologically, the same as its projection onto the x-axis, which in turn is the x-axis with a round hole of radius $|\tau|$ around the origin. In the same way, the domain $|\mu| \geq 1$ may be regarded as the y-axis with a similar hole. The boundaries of the two holes are glued together in the curve $xy = \varepsilon'$ which is the same as saying that the two planes are joined by a tube.

Part 3

Euler Groups and Arithmetic of Geometric Progressions

Euler Groups and Arithmetic of Geometric Progressions

1. Basic Definitions

For any natural number n, the set $\mathbb{Z}_n = \mathbb{Z}/n\mathbb{Z}$ of residues modulo n contains a multiplicative group $\Gamma(n) \subset \mathbb{Z}_n$, formed by the residues relatively prime to n.

Definition. The *Euler group* $\Gamma(n)$ is the multiplicative group of residues modulo n which are relatively prime to n.

Gauss called the number of elements $\varphi(n)$ of the group $\Gamma(n)$ the value of the *Euler function* φ.

Thus the Euler group is a commutative group of order $\varphi(n)$. Many have researched the Euler function (Fermat, Euler, Gauss, Legendre, Jacobi and others). But the Euler group is much more interesting than the number $\varphi(n)$ given by the Euler function, just as the homology groups are more interesting than the Betti numbers.

Reduction modulo a defines a natural homomorphism $\Gamma(ab) \to \Gamma(a)$. The present work is dedicated to a description of the Euler group and these natural homomorphisms.

Remark. I will not dwell on the question of who was the first to discover this or that fact described below. But one can find in the literature (see [**2**], [**22**], [**20**], [**25**], [**34**]) descriptions in various terms such as: "This result was known to Fermat, was formulated by Euler, and was proven by Gauss (the proofs were then refined by so-and-so)". I prefer to consider what follows as a worthy of inclusion in elementary textbook exposition of "Euler theory", without worrying about the absence in his published works either of formulations or of proofs.

2. A Digression on the Euler Function

It is easy to compute the values of the Euler function using the prime factorization of the argument, $n = p_1^{a_1} \cdots p_k^{a_k}$; namely:

$$\varphi(n) = (p_1 - 1)p_1^{a_1-1} \cdots (p_k - 1)p_k^{a_k-1}.$$

For example, $\varphi(p) = p-1, \varphi(9) = 6, \varphi(15) = 8$ (and, by definition, $\varphi(1) = 1$).

Indeed, every residue other than 0 modulo the prime p is relatively prime to p, so $\varphi(p) = p - 1$.

Of the p^a residues modulo $n = p^a$, the residues which are not relatively prime to n are just those which are divisible by p, of which there are p^{a-1}. So $\varphi(p^a) = p^a - p^{a-1}$.

Finally, if $p_1, \ldots, p_k$ are all prime factors of n, then a remainder modulo n which is relatively prime to n has a remainder r_i modulo $p_i^{a_i}$ which is relatively prime to p_i, and is uniquely determined by these remainders r_i. (For a formal proof see Section 6, where this follows from Theorem 1.)

For large values of the argument n, the value of $\varphi(n)$ grows, on the average, as cn, where $c = 6/\pi^2$, which is close to $2/3$ (see [3]). The "average growth" referred to in [3] is defined by the condition that the limit as $n \to \infty$ of the ratio of the sum of the first n values is equal to 1:

$$\lim_{n\to\infty} \frac{\varphi(1) + \varphi(2) + \cdots + \varphi(n)}{c \cdot 1 + c \cdot 2 + \cdots + cn} = 1.$$

This does not exclude rather large differences between certain values of $\varphi(n)$ and cn. All it means is that these are rare.

The constant c is the probability that the fraction x/y, with integers x and y, be in lowest terms. It is defined as the limit as $R \to \infty$ of the ratio of the number of uncancellable pairs (x, y) in the disk $x^2 + y^2 \leq R^2$ to the number of all such pairs (which grows as πR^2 as R increases).

This probability was computed by Gauss and the result published by Dirichlet [**19**]. For the analogous problem about vectors in $\mathbb{Z}^m$ the probability of uncancellability is equal to $c = 1/\zeta(m)$, where *Euler's zeta function* is defined as the sum of the series

$$\zeta(m) = \frac{1}{1^m} + \frac{1}{2^m} + \frac{1}{3^m} + \cdots .$$

The proof of the formula for c is as follows. The probability of cancellability by 2 is equal to $1/2^m$ (since each of the m components must be divisible by 2). The probability of cancellability by the prime p is $1/p^m$, and the probability of uncancellability by p is $1 - 1/p^m$.

Cancellabilities by various prime numbers p are clearly independent, so the probability of complete uncancellability is equal to $c = \prod \frac{1}{1 - \frac{1}{p^m}}$ (where the product is taken over all the primes). But the uniqueness of the prime factorization of the number n implies the well-known formula of Euler

$$\prod_p \frac{1}{1 - \frac{1}{p^m}} = \sum_n \frac{1}{n^m}$$

which laid out the foundation of the theory of graded algebras.

Euler's formula follows from the expression for the sum of a geometric progression,

$$\frac{1}{1-\frac{1}{p^m}} = 1 + \frac{1}{p^m} + \frac{1}{p^{2m}} + \cdots,$$

because of the uniqueness of the decomposition of n into prime factors.

Finally, the formula $\zeta(2) = \pi^2/6$ for this value of the zeta function follows from the theory of Fourier series. Namely, consider the 2π-periodic extension f of the function $|t| - \pi/2$, defined on the interval $|t| \leq \pi$. The Fourier coefficients are easy to compute (and decrease as $1/n^2$). The expression $f(0) = -\pi/2$ in terms of these coefficients gives the value $\pi^2/6$ for $\sum 1/n^2$ (see below).

Thus an investigation of the growth of the Euler function φ includes all of mathematics, from Fourier series to probability theory and the theory of graded algebras.

The function f, which we met in computing the value $\zeta(2)$, turns out to be a member of Kolmogorov's famous sequence of periodic functions which starts with the function $F_0 = \operatorname{sgn}(\cos t)$ and continues according to the rule $F'_{n+1} = F_n$.

These functions approximate the sine and cosine with piecewise-polynomial functions of increasing degree, from the step function F_0 and the sawtooth function $F_1 = f$ to the parabolically approximating continuously differentiable function F_2 and the n times continuously differentiable function F_{n+1}. Kolmogorov invented them in order to solve a remarkable extremal problem: find the greatest value of the intermediate k-th derivative of a 2π-periodic function with given upper bounds for the absolute value of the function and its highest (m-th) derivative.

His estimate is suggested by dimension theory and by Leonardo da Vinci's self-similarity principle, taking into account the dimension of the derivative as expressed in Leibniz' notation:

$$\dim \frac{d^r y}{(dx)^r} = \frac{\dim y}{(\dim x)^r}.$$

Kolmogorov's estimate has the form

$$\left\| \frac{d^k y}{(dx)^k} \right\| \leq C \|y\|^a \left\| \frac{d^m y}{(dx)^m} \right\|^b,$$

where the rational exponents a and b are equal to $b = k/m, a = 1 - b$, by the self-similarity principle. The constant C is achieved at the appropriate function of Kolmogorov's sequence. (And if the period T differs from 2π, then the similarity arguments also dictate the form of the dependence of the constant C on T.)

For example, the first derivative is approximated by the square root of the product of the maxima of absolute values of the function and its second derivative. This particular case of Kolmogorov's theorem was established earlier by Hadamard and Littlewood, independently of each other.

The general case of Kolmogorov's inequality is essentially the first result of the contemporary theory of controlled dynamical systems (which became widely known much later, when Pontryagin published his "maximum principle"). Kolmogorov's proof, which is based on Huygens' general geometric principle of wave propagation (the "maximum principle" is also a variant of the latter), can be applied, with small changes, to the general theory of controlled systems (just as the solution of the brachistochrone problem contains, in essence, all of the calculus of variations). The main idea here is the transition from Huygens' principle of enveloping wave fronts to its infinitesimal variant, which is the system of canonical Hamiltonian equations in the phase space.

The practical implication from these general ideas consists in the principle that to get an optimization in problems of control with bounds, we must always give an extreme value to the control (in Kolmogorov's problem this is the highest derivative). For example, the second derivative in the situation of Hadamard and Littlewood must always take either a maximal or minimal value, which immediately leads (taking into account periodicity) to extrema which are proportional to F_2 in this problem. This is how Kolmogorov arrived at his sequence of functions F_n.

As for the Fourier series of the even function $f = F_1$ (which we need for the computation $\zeta(2) = \pi^2/6$), we can find then from the formula representing f in this series, $f(t) = \sum a_k \cos(kt)$. For $k = 0$ we have $a_0 = \int_{-\pi}^{\pi} f(t)\,dt = 0$. For $k \neq 0$ we can use integration by parts:

$$\begin{aligned}\pi a_k &= \int_{-\pi}^{\pi} f(t)\cos(kt)dt = 2\int_0^{\pi} t\cos(kt)dt \\ &= 2\left[\left(\frac{kt}{k^2}\sin(kt)\right)\Big|_0^{\pi} - \frac{1}{k^2}\int_0^{\pi}\sin(kt)d(kt)\right] \\ &= \frac{2}{k^2}(\cos(kt))\Big|_0^{\pi} = \begin{cases} 0, & \text{if } k \text{ is even,} \\ -\dfrac{4}{k^2}, & \text{if } k \text{ is odd.}\end{cases}\end{aligned}$$

Thus the Fourier coefficients with even subscripts are all zeroes, and for odd k

$$a_k = \left(-\frac{4}{\pi}\right)\frac{1}{k^2}.$$

Therefore the value $f(0) = -\dfrac{\pi}{2}$ is equal to the sum of the Fourier series

$$f(0) = \left(\sum_{m=0}^{\infty}\frac{1}{(2m+1)^2}\right)\left(-\frac{4}{\pi}\right),$$

and we have computed the sum of the series of inverse squares of odd numbers

$$A = \sum_{m=0}^{\infty}\frac{1}{(2m+1)^2} = \frac{\pi^2}{8}.$$

Let us introduce the notation B for the required sum of the series of inverse squares for all the natural numbers. Since each natural number is either odd or even, we have:

$$\sum_{k=1}^{\infty}\frac{1}{k^2}=\sum_{m=1}^{\infty}\frac{1}{(2m+1)^2}+\sum_{m=1}^{\infty}\frac{1}{(2m)^2}=\sum_{m=1}^{\infty}\frac{1}{(2m+1)^2}+\frac{1}{4}\sum_{m=1}^{\infty}\frac{1}{m^2};$$

that is, $B = A + B/4$. It follows (since we already know the odd part, $A = \pi^2/8$, from the Fourier series), that

$$\zeta(2) = B = \frac{4}{3}A = \frac{\pi^2}{6}.$$

3. Tables for Euler Groups

Direct computation of $\Gamma(n)$ for the Euler group gives the values in the following table for the first few values of n. The notation of the form $2^a.3^b.4^c$ in this table represents a commutative group, isomorphic to the group $(\mathbb{Z}_2)^a\times(\mathbb{Z}_3)^b\times(\mathbb{Z}_3)^c$ (whose order is equal to the product $\varphi = 2^a3^b4^c$).

Of course, the group $2^a.3^a$ is the same group as 6^a, but the group $2^a.4^a$ is different from the group 8^a, and the group 2^{2a} is different from the group 4^a.

Here is a table for the first few Euler groups.

n	3	4	5	6	7	8	9	10	11	12	13	14	15
$\Gamma(n)$	2	2	4	2	6	2^2	6	4	10	2^2	12	6	$4\cdot 2$
g_i	2	3	2 3	5	3 5	$(3,5)$	2 5	3 7	2, 7 6, 8	$(5,7)$	2, 6 7,11	3 5	$(2,11)$

n	16	17	18	19	20	21	22	23
$\Gamma(n)$	$4\cdot 2$	16	6	18	$4\cdot 2$	$6\cdot 2$	10	22
g_i	$(3,7)$	3,5,10,11 6,7,12,14	5 11	2, 3, 14 10,13,15	$(3,11)$	$(2,5)$	7, 13 19,17	5, 7, 11,15, 17 14,10,21,20, 19

n	24	25	26	27	28	29
$\Gamma(n)$	2^3	20	12	18	$6\cdot 2 = 2^2\cdot 3$	28
g_i	$(5,7,13)$	2, 3, 8, 12 13,17,22,23	7, 11 15,19	2, 5,11 14,20,23	$(3,13),(13,27,9)$	2, 3, 8, 14,18,19 15,10,11,27,21,26

n	30	31	32	33	34	35
$\Gamma(n)$	$4 \cdot 2$	30	$8 \cdot 2$	$10 \cdot 2 = 2^2 \cdot 5$	16	$12 \cdot 2$
g_i	$(7, 11)$	3, 11, 12, 22 21, 17, 13, 24	$(3, 15)$	$(2, 10), (10, 32, 4)$	3, 5, 11, 27 23, 7, 31, 29	$(2, 6)$

The numbers g_i given in the third row of the table (for cyclic groups $\Gamma(n) \subset \mathbb{Z}_n$) are the cyclic generators of the indicated cyclic group. That is, the numbers g^k $(0 \leq k < \varphi(n))$ give us the whole group $\Gamma(n)$. Also, under every generator g the inverse generator h is shown (so that $gh \equiv 1 \pmod{n}$). The cyclic generators of the group $\Gamma(n)$ are also called the *primitive roots modulo n*.

For non-cyclic groups, in the third row we indicate in parentheses a possible choice of cyclic generators of the group-factors. (Other such choices are easily obtained by exchanging these generators with their powers and products.)

The proofs of the theorem shown in the table can be obtained by direct computation of geometric progressions $a^k \pmod{n}$. To identify non-cyclic Euler groups it is convenient to construct an oriented graph, whose vertices are elements of the group, with arrows leading to the squares of the elements (in additive notation – from x to $2x$).

Example. The graphs of the groups of order 8 are as follows:

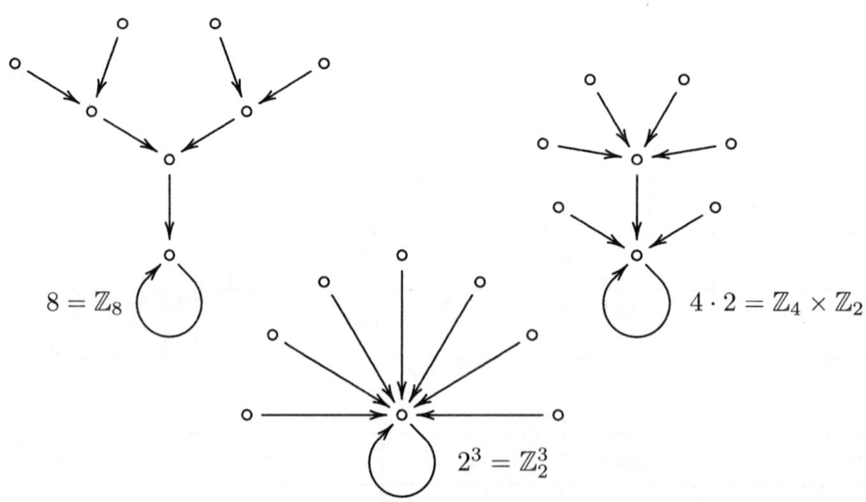

It is also useful to construct a table of the number of elements of different orders in the group. For the previous example the table is as follows:

group \ order	1	2	4	8
$\mathbb{Z}_8$	1	1	2	4
$\mathbb{Z}_4 \times \mathbb{Z}_2$	1	3	4	0
$\mathbb{Z}_2^3$	1	7	0	0

4. Euler Groups of Products

An analysis of the table in Section 3 immediately leads to the following conclusions (whose proofs we discuss below in Section 6).

Theorem 1. *If the numbers a and b are relatively prime, then the Euler group of their product is the direct product of the two Euler groups: $\Gamma(ab) = \Gamma(a) \times \Gamma(b)$.*

Theorem 2. *If the number $n = p$ is prime, then its Euler group is cyclic, $\Gamma(p) = \mathbb{Z}_{p-1}$.*

Theorem 3. *If the number $n = p^a$ is a power of an odd prime, then its Euler group is cyclic,*

$$\Gamma(p^a) = \mathbb{Z}_{\varphi(p^a)} = \mathbb{Z}_{(p-1)p^{a-1}}.$$

Theorem 4. *If the number $n = 2^a$, $a > 2$ is a power of two, then its Euler group is the product of cyclic groups of orders 2 and 2^{a-2}:*

$$\Gamma(2^a) = \mathbb{Z}_2 \times \mathbb{Z}_{2^{a-2}}.$$

These four theorems describe all the Euler groups, since every integer can be factored into primes.

Corollary. *The Euler group $\Gamma(n)$ is cyclic if and only if the number n is equal to 2, or 4, or a power of an odd prime, or twice a power of an odd prime.*

Theorem 2 is simply Fermat's "Little Theorem", as generalized by Euler (primitivity), and by Gauss.

5. The Homomorphism Given by Reduction Modulo a, $\Gamma(ab) \to \Gamma(a)$

Let us denote the reduction of numbers from modulo ab to modulo a by $\pi : \mathbb{Z}_{ab} \to \mathbb{Z}_a$. (We will sometimes use the same symbol for the restriction of this reduction to $\Gamma(ab) \to \Gamma(a)$, or for its action on $\mathbb{Z}$.)

Remark. $\pi(\Gamma(ab)) \subset \Gamma(a)$ for the the following reason: if the residue of x (mod a) were not relatively prime to a, then the residue of x (mod ab) would not be relatively prime to ab.

As we will now prove, $\pi(\Gamma(ab)) = \Gamma(a)$ (although this is not quite obvious).

Let the number x be relatively prime to a; that is, $(x, a) = 1$. Let us find the residue modulo ab, relatively prime to ab and projected onto the residue $x \pmod a$. We must study all the pre-images of the residue $x \pmod a$ in $\mathbb{Z}_{ab}$ and find among them an element of $\Gamma(ab)$. First, let us prove a slightly more general version of the "Chinese Remainder Theorem".

Theorem $\mathbf{T_{D,B}}$. *In the arithmetic progression $\{x+nD\}$, $(n = 0, 1, \dots)$ where the initial term x and the difference D are relatively prime (so that $(x, D) = 1$) there exists an element relatively prime to B:*

$$\forall B\ \exists n\colon (x + nD, B) = 1.$$

Proof. Theorem $T_{1,B}$ is obvious. We now assume that Theorems $T_{d,b}$, with $d < D$ are true, and deduce from them Theorem $T_{D,B}$. So suppose that $(x, D) = 1$.

We denote by δ the greatest common divisor of the numbers B and D, so that

$$B = \beta\delta,\ D = \gamma\delta,\ (\beta, \gamma) = 1.$$

First Case. $\gamma = 1, D = \delta$.

In this case, $B = \beta D > D$, or $\beta = 1$, $B = D$. If $B = D$, then $(x, B) = 1$ from the given conditions, so the choice $n = 0$ furnishes a proof of theorem $T_{D,D}$.

In the case $\beta > 1$, when $D < B$, we have for $D > 1$ (when $\beta < B$) the implication

$$T_{D,\beta} \to T_{D,B}\ ,$$

since the fact that $x+nD$ is relatively prime to β implies that it is relatively prime to $B = \beta\delta$ (the number $x + nD$ is relatively prime to δ, as is x, by the hypothesis of Theorem $T_{D,B}$ (where $D = \delta$), which we are proving).

Thus Theorem $T_{D,B}$ reduces to theorems with the same D, but with smaller values of B (and as long as $B < D$, the condition $\gamma = 1$ is not fulfilled).

Second Case. $\gamma > 1$, $\delta < D$.

In this case we can deduce theorem $T_{D,B}$ from Theorem $T_{\delta,\beta}$, where $D = \gamma\delta$, $B = \beta\delta$, $(\beta, \gamma) = 1$.

The condition $(x, D) = 1$ implies that the number x is relatively prime to any divisor δ of the number D. By Theorem $T_{\delta,\beta}$, there exists an integer m such that the number $r = x + m\delta$ is relatively prime to β.

From the fact that β and γ are relatively prime, it follows that we can write $p\beta + q\gamma = 1$ (for some integers p and q). So now we can express r in the form

$$r = x + \delta(mp\beta + mq\gamma) = x + mpB + nD,$$

where $n = mq$.

Consider the number

$$R = x + nD = r - mpB,$$

which will prove the theorem for us. This number is relatively prime to β, since r is relatively prime to β (by Theorem $T_{\delta,\beta}$ and our choice of m), while the number $B = \beta\delta$ is divisible by δ. Also, R is relatively prime to δ, since $(x, \delta) = 1$, by the assumption of Theorem $T_{D,B}$, while the number $D = \gamma\delta$ is divisible by δ.

Thus, the number R must be relatively prime to the product $\beta\delta = B$, which proves Theorem $T_{D,B}$ (if we take $-mq$ for n).

So Theorem $T_{D,B}$ is now proved for any D and B.

The equality $\pi(\Gamma(ab)) = \Gamma(a)$ follows from these theorems, since by Theorem $T_{a,b}$, among the residues of the numbers $x + na$, $n = 0, 1, 2, \ldots$ (mod b), there must be a residue which is relatively prime to b (if $(x, a) = 1$; that is, if $x \in \Gamma(a)$).

Of course, the number n can be taken in the interval $\{0, 1, \ldots, b-1\}$, since if n is increased by b the residue of the number $x + na$ (mod ab) stays unchanged.

Corollary. *The number of pre-images of a point $x \in \Gamma(a)$ under the mapping $\pi\colon \Gamma(ab) \to \Gamma(a)$ is the same for all x.*

Proof. As we've just proved, the mapping π is a homomorphism of the group $\Gamma(ab)$ onto the group $\Gamma(a)$. Thus the number of pre-images referred to in the statement is equal to the order of the kernel of this homomorphism, $|\operatorname{Ke}\pi| = \varphi(ab)/\varphi(a)$.

6. Proofs of the Theorems on Euler Groups

Proof of Theorem 1. Let us compare the two homomorphisms of reduction modulo a and modulo b:

$$\pi : \Gamma(ab) \to \Gamma(a), \ \ \rho : \Gamma(ab) \to \Gamma(b).$$

Let $x \in \Gamma(a)$, $\pi^{-1}(x) = \{x + na\}, 0 \le n < b$.

The b residues of the numbers $x+na$ modulo b are all distinct: otherwise the number $n_1a - n_2a$ would be divisible by b for $|n_1 - n_2| < b$, which contradicts the assumption in Theorem 1 that a and b are relatively prime.

This means that there is an element in $\mathbb{Z}_{ab}$ with residue x modulo a and with any residue modulo b. It also follows that the mapping

$$\pi \times \rho : \Gamma(ab) \to \Gamma(a) \times \Gamma(b)$$

covers the entire product (the element z of $\mathbb{Z}_{ab}$ which is congruent to $x \in \Gamma(a)$ (mod a) and to $y \in \Gamma(b)$ (mod b) must lie in $\Gamma(ab)$, since the fact that z is relatively prime to both a and b implies that it is also relatively prime to the product ab.

On the other hand, *the kernel of the mapping $\pi \times \rho$ is trivial*, since its elements are congruent to 1, both modulo a and modulo b. This shows that the difference of any two of them in $\mathbb{Z}_{ab}$ is divisible by the product ab (since a and b are relatively prime by the assumption in Theorem 1). That is, this difference is zero.

Theorem 2 is Fermat's Little Theorem, supplemented by the existence of primitive roots modulo a prime (which we prove below, in Section 12). The existence of primitive roots is needed to show that the order of the group does not turn out to be smaller than $p-1$.

Proof of Theorem 3. For an odd prime p, we can write the number p^{a+1} in the form $p \cdot p^a$, and look at the corresponding group homomorphism induced by the mapping π of reduction modulo p:

$$\pi : \Gamma(p^{a+1}) \to \Gamma(p).$$

The image is the entire group $\Gamma(p)$, which is cyclic with order $p-1$ (by Theorem 2). Let us study the kernel of the homomorphism π.

Lemma. *The kernel of the homomorphism π is a cyclic group of order p^a.*

Proof of Lemma. The kernel K of π has order $\varphi(p^{a+1})/\varphi(p) = p^a(p-1)/(p-1) = p^a$. It consists of those elements of $\Gamma(p^{a+1})$ which are congruent to 1 modulo p. In particular, $1+p \in K$. We will show that if $0 < k < p^a$, then $(1+p)^k \not\equiv 1 \pmod{p^{a+1}}$. That is, the cyclic subgroup of K generated by $1+p$ cannot have order less than p^a, so K is a cyclic group of order p^a (generated by $1+p$).

If $k < p^a$, then $k = \ell p^b$ where $b < a$ and ℓ is not divisible by p. We use the binomial formula:

$$\begin{aligned}(1+p)^k = (1+p)^{\ell p^b} = 1 \quad &+ \ell p^b \cdot p + \frac{\ell p^b(\ell p^b - 1)}{2} \cdot p^2 + \cdots \\ &+ \frac{\ell p^b(\ell p^b-1)\cdots(\ell p^b - n + 1)}{n!} \cdot p^n + \cdots .\end{aligned}$$

Let us show that if $n \geq 2$, then

$$\frac{\ell p^b(\ell p^b-1)\cdots(\ell p^b - n + 1)}{n!} \cdot p^n$$

is divisible by p^{b+2}. We will deduce this from the fact that if p^m is the maximal power of p which divide $n!$ $(n \geq 2)$, then $m \leq n-2$. Indeed, the product $1 \cdot 2 \cdot \ldots \cdot n = n!$ contains $\left[\frac{n}{p}\right]$ factors divisible by p, $\left[\frac{n}{p^2}\right]$ factors divisible by p^2, $\left[\frac{n}{p^3}\right]$ factors divisible by p^3, and so on. Hence

$$\begin{aligned} m &= \left[\frac{n}{p}\right] + \left[\frac{n}{p^2}\right] + \left[\frac{n}{p^3}\right] + \cdots \\ &< \frac{n}{p} + \frac{n}{p^2} + \frac{n}{p^3} + \cdots = \frac{n}{p}\frac{1}{1-\frac{1}{p}} = \frac{n}{p-1} \leq \frac{n}{2} \leq n-1, \end{aligned}$$

if $n \geq 2$. We see that $m < n-1$, so $m \leq n-2$. This gives us the desired result: the product $\ell p^b(\ell p^b - 1)\cdots(\ell p^b - n+1)\cdot p^n$ is divisible by p^{b+n}, while the prime factorization of $n!$ contains at most $n-2$ factors p. Thus, our fraction is divisible by p^{b+2}.

Hence, in the right hand side of the binomial formula given above, all summands, except 1 and $\ell p^b \cdot p = \ell p^{b+1}$ are divisible by p^{b+2}, so

$$\begin{aligned} (1+p)^k &\equiv 1 + \ell p^{b+1} && \pmod{p^{b+2}}, \\ (1+p)^k &\not\equiv 1 && \pmod{p^{b+2}}, \\ (1+p)^k &\not\equiv 1 && \pmod{p^{a+1}} \end{aligned}$$

(the last is true because $b < a$). This completes the proof of Lemma.

End of Proof of Theorem 3. The group $\Gamma(p^{a+1})$ is an Abelian group of order $p^a(p-1)$. According to a theorem of group theory, it must be a product of cyclic groups whose orders are powers of primes. But since $\Gamma(p^{a+1})$ contains a cyclic subgroup K of order p^a, this product must be $K \times L$ where $|L| = p - 1$. Moreover

$$L = \Gamma(p^a)/K = \Gamma(p^a)/\operatorname{Ke}\pi = \Gamma(p) \cong \mathbb{Z}_{p-1}$$

(by Theorem 2), so

$$\Gamma(p^{a+1}) \cong \mathbb{Z}_{p^a} \times \mathbb{Z}_{p-1} \cong \mathbb{Z}_{(p-1)p^a},$$

as stated in Theorem 3.

Proof of Theorem 4. Consider the homomorphism of reduction modulo 4, $\pi\colon \Gamma(2^a) \to \Gamma(4)$. The image consists of residues 1 and 3 $\pmod 4$, and we can write each element of the group $\Gamma(2^a)$ as the residue of the number

$$x = 1 + 2\alpha + 4u,\ 0 \le \alpha,\ u < 2^{a-2}.$$

In particular, there is one special element of order two

$$w = 2^{a-1} - 1 = 1 + 2\alpha + u,\ \alpha = 2^{a-3} - 1,\ u = 0.$$

For this element we have $w^2 = 1 \pmod{2^a}$, since the numbers 2^{2a-2} and $2 \cdot 2^{a-1}$ are divisible by 2^a for $a \ge 2$ (we assumed in Theorem 4 that $a \ge 2$). We obtain the subgroup $\{1, w\}$ of $\Gamma(2^a)$.

Any element x in $\Gamma(2^a)$ can be uniquely written in either the form $x = 1 + 4u$ (when $\pi(x) = 1$) or $x = w(1+4z)$ (when $\pi(x) = 3$, as well as $\pi(w)$). Indeed, $x = w \cdot wx$, and if $\pi(x) = 3$, then $\pi(wx) = \pi(w)\pi(x) = 9 = 1 \pmod 4$, so $wx = 1 + 4z$ for some z.

Thus we have presented the group $\Gamma(2^a)$ in the form of a direct product of non-intersecting subgroups $\mathbb{Z}_2 = \{1, w\}$ and $\{1+4z\}$, where $0 \le z < 2^{a-2}$. Theorem 4 is then implied by the following fact.

Lemma. *The group* $\{1 + 4z\}$ *(where* $0 \le z < 2^{a-2}$*) of residues modulo* 2^a *is cyclic:* $\{1 + 4z\} \approx \mathbb{Z}_{2^{a-2}}$.

Proof of Lemma. In the same way that we proved Theorem 3 above, we will prove that the element $1 + 4 = 5$ is a cyclic generator, which we can write in the form

$$q_0 = 1 + 4 + 8D_0,\ D_0 = 0.$$

Now suppose we are given the element $Q = 1 + 2^{2+i} + 2^{3+i}D$ in $\Gamma(2^a)$. A computation of its square shows that

$$\begin{aligned} Q^2 &= 1 + 2 \cdot 2^{2+i} + 2 \cdot 2^{3+i}D + 2^{2(2+i)} + 2^{2(3+i)}D^2 + 2 \cdot 2^{2+i}2^{3+i}D \\ &= 1 + 2^{2+(i+1)} + 2^{3+(i+1)}D'. \end{aligned}$$

(The number D' is an integer because the powers of two in the last four terms of the squared polynomial are never less than those in the the coefficient of D' as written: $2(2+i) \geq 3+(i+1)$, $2(3+i) > 3+(i+1)$, $1+(2+i)+(3+i) > 3+(i+1)$.)

Applying this filtered squaring formula to the case $Q = q_0$ (where $i = 0, D = 0$), we get (for $q_1 = Q^2 = q_0^2$) the expression

$$q_1 = 1 + 2^{2+1} + 2^{3+1}D_1.$$

Applying our formula to the case $Q = q_1$, we find, for $q_2 = q_0^4 = q_1^2$, the expression

$$q_2 = 1 + 2^{2+2} + 2^{3+2}D_2.$$

If we continue squaring, we obtain the sequence

$$q_g = q_0^{2^g} = q_{g-1}^2 = 1 + 2^{2+g} + 2^{3+g}D_g$$

for $g = 0, 1, 2, \ldots, a-3$. For $g = a-2$ we obtain $q_{a-2} = q_0^{2^{a-2}} = 1 + 2^a + 2^{a+1}D_{g-2} \equiv 1 \pmod{2^a}$; that is, the number $5^{2^{a-2}} \equiv 1 \pmod{2^a}$ is an identity element for the group $\Gamma(2^a)$.

And we can show that all the previous powers of 5 are different from 1 modulo 2^a. Indeed, let us express the binary expansion of the number N which is less than 2^{a-2} as

$$N = N_0 + N_1 \cdot 2 + N_2 \cdot 2^2 + \cdots + N_{a-3} \cdot 2^{a-3},$$

where each N_i is either 0 or 1.

Then for 5^N we obtain the expression

$$q_0^N = q_0^{N_0} q_1^{N_1} q_2^{N_2} \cdots q_{a-3}^{N_{a-3}}.$$

Let i be the number of the first non-zero coefficient $N_0, N_1, \ldots, N_{a-3}$ of the binary expansion of N. From the formula for q_g proved above, we have

$$q_0^N \equiv 1 + 2^{2+i} \pmod{2^{3+1}},$$

so $q_0^N \not\equiv 1$ in $\Gamma(2^a)$ (since $i \leq a-3$).

This shows that all of the 2^{a-2} elements of the group $\{1+4z\} \pmod{2^a}$ of the form $5^N \pmod{2^a}$ are distinct, so that this subgroup $\{1+4z\}$, consisting of 2^{a-2} elements in all, is exhausted by the elements of the form 5^N and must be cyclic.

The lemma is proved, and this concludes the proof of Theorem 4.

7. Fermat-Euler Dynamical Systems

Let us fix a number a relatively prime to n, and consider multiplication by a as a transformation A of the set $\Gamma(n)$ of residues modulo n relatively prime to n into itself. This transformation takes the residue of the number x into the residue of the number ax (which is relatively prime to n, just as x is). This defines a permutation $A\colon \Gamma(n) \to \Gamma(n)$, $x \mapsto ax$.

The transformation A of the set $\Gamma(n)$ onto itself, like any one-to-one transformation, can be decomposed into cycles of these permutations of the $\varphi(n)$ elements.

Theorem (Euler-Fermat). *All the cycles of the Fermat-Euler permutation $A\colon \Gamma(n) \to \Gamma(n)$ have the same period $T(n, a)$.*

Proof. Any element y of the group $\Gamma(n)$ can be obtained from any other its element x by multiplying on the right by some element z. Therefore

$$A^T y = A^T(xz) = (A^T x)z = xz = y;$$

that is, the period T of the element x is also a period of y.

Corollary. *The set $\Gamma(n)$ of residues modulo n which are relatively prime to n can be decomposed into $N(n,a) = \varphi(n)/T(n,a)$ non-intersecting orbits of the Fermat-Euler transformation. Thus the number of orbits of N, as well as the period T, divides the value of the Euler function $\varphi(n) = NT$, and satisfy the Fermat-Euler equation*

$$a^{\varphi(n)/N} \equiv 1 \pmod{n}.$$

Example. If the number $n = p$ is prime, then we have the equation

$$a^{p-1} \equiv 1 \pmod{p}.$$

This again proves Fermat's original "Little Theorem".

Euler applied this theorem to composite numbers n in place of p, noticing that the exponent $p-1 = \varphi(p)$ must be replaced by $\varphi(n)$. This is the origin of the Euler function φ.

If the Fermat-Euler transformation has only one orbit ($N = 1$), then its period is $T = \varphi(n)$, so that the Fermat-Euler equation assumes its simplest form

$$a^{\varphi(n)} \equiv 1 \pmod{n},$$

which is true even when there are more orbits.

The questions of the behavior of the period, the number of orbits, and their dependence on n are not at all simple. Below we talk about (mostly experimental) data, obtained by means of computation of the values of the functions $N(n)$ and $T(n)$ for the simplest case $a = 2$ (assuming that n is odd, i.e. relatively prime to the base a).

The values of the number of orbits $N(n)$ and the period $T(n)$ of the operation A of multiplication by $a = 2$ for the first five decades of odd moduli n are shown in the following table:

n	**3**	**5**	**7**	9	**11**	**13**	15	**17**	**19**	21	**23**	25	27	**29**	**31**	33	35	**37**	39
N	1	1	2	1	1	1	2	2	1	2	2	1	1	1	6	2	2	1	2
T	2	4	3	6	10	12	4	8	18	6	11	20	18	28	5	10	12	36	12

n	**41**	**43**	45	**47**	49	51	**53**	55	57	**59**	**61**	63	65	**67**	69	**71**	**73**	75	77
N	2	3	2	2	2	4	1	2	2	1	1	6	4	1	2	2	8	2	2
T	20	14	12	23	21	8	52	20	18	58	60	6	12	66	22	35	9	20	30

n	**79**	81	**83**	85	87	**89**	91	93	95	**97**	99
N	2	1	1	8	2	8	6	6	2	2	2
T	39	54	82	8	28	11	12	10	36	48	30

The prime numbers n are given here in boldface. For each of them, $NT = n - 1$. In other cases the product $NT = \varphi(n)$ is smaller.

The Euler-Fermat theorem implies that the Young diagram describing the decomposition of the set $\Gamma(n)$ into orbits by the Fermat-Euler transformation A is always a rectangle of area $\varphi(n)$, with its base of length $T(n)$ and $N(n)$ rows of length $T(n)$.

Each row is an orbit of the action of the transposition A, and we can list its elements in the order $(x, Ax, A^2x, \dots)$. For $n = 15$ this Young diagram is as follows:

7	14	13	11	$(11 \cdot 2 \equiv 7)$
1	2	4	8	$(8 \cdot 2 \equiv 1)$

8. Statistics of Geometric Progressions

The growth of the function $T(n)$ as the modulus n increases appears rather irregular, so that the Young diagrams are sometimes very tall (with a relatively large ratio N/T, such as $N = 48, T = 9$ for $n = 511$) and sometimes low (with a small ratio N/T, such as $1/82$ for $n = 83$).

The sums for the decades of values $1 \le n \le 19$, $21 \le n \le 39$, $41 \le n \le 59$, $61 \le n \le 79$, $81 \le n \le 99$ and for the corresponding values of the periods T show a more regular dependence:

$\sum n$	100	300	500	700	900
$\sum T$	68	158	246	299	329

These data hint at a linear rate of growth for T as a function of n (roughly speaking, on the average, $T = Cn$, where $C \approx \dfrac{3}{7}$ for $n \sim 70$, and with slowly decreasing values as n grows).

I do not know any theorems regarding this matter. Here, however, are a few ideas of a non-mathematical nature.

The errors in the value of x increase as x is multiplied by a very large number a^t (for large values of the time t), caused during the time t, by the dynamical system of Fermat-Euler $A\colon \Gamma(n) \to \Gamma(n)$. This observation leads one to the idea that the orbit $\{a^t x,\ t = 1, \dots, T\}$ must be scattered within $\Gamma(n)$ (and even within $\mathbb{Z}_n = \mathbb{Z}/n\mathbb{Z}$) in a chaotic fashion.

The length or the period T of this orbit is defined by the condition that all T "random points" $a^t x$ must occupy different positions within the $m = \varphi(n)$ points of the set $\Gamma(n)$ (or within all the $m = n$ points of the set $\mathbb{Z}_n$). Therefore the idea presents itself to compare $T(n)$ with the length of a typical random sequence of independent choices of one of m points of some set, in which all the elements of this random sequence remain distinct.

This problem in probability theory is usually called the "birthday problem" (T is the number of students in some group, $m = 365$ is the number of possibilities for their birthdays). The question is: what is the probability that the birthdays of all T students in the group are different?

Clearly, this probability gets smaller as the number of students T gets larger, and gets very small starting with a certain number of students (it is even equal to 0 for $T > m$). On the other hand, if the number T of students is small, then the probability of coincidence of birthdays is small.

The critical size of the group T_*, where the probability of a coincidence is small for $T \ll T_*$ and large for $T \gg T_*$, increases as the square root of m.

The easiest way to convince oneself of that is to divide the number of non-overlapping choices of T elements of an m-element set (equal to the product of T factors $m(m-1)\cdots(m-T+1)$) by the number of all possible choices (which is equal to m^T):

$$p = 1\left(1 - \frac{1}{m}\right)\left(1 - \frac{2}{m}\right)\cdots\left(1 - \frac{T-1}{m}\right).$$

If we suppose that the number T is small relative to m (for example, if m goes to infinity for a fixed T), we obtain (neglecting the details of an estimate of the error) $\ln p \approx -\sum\limits_{j=1}^{T-1} j/m$.

Thus an approximate formula for the probability of no coincidence is

$$p \approx e^{-T(T-1)/2m}.$$

This number is small when $T \gg \sqrt{2m}$, and close to 1 when $T \ll \sqrt{2m}$, so that the length of the segment of the m-valued random sequence without coincidences grows as the square root of the number of values.

Remark. I think that the transition from small values of p to large values as T passes through the critical value which is of order $\sqrt{m}$ is described by the universal function *erf* (the integral of the Gaussian density) if we use appropriate units of measurement (depending on m) of the deviation of the value of T from the critical value. Unfortunately, I have not seen in the literature a proof of the corresponding theorem, despite its obvious importance for many applications. This question, apparently, is too simple for probabilists to look into.

Comparing this fact from the theory of probability with our table of values $T(n)$, we come to the conclusion that the observed periods $T(n)$ of geometric progressions 2^t grow faster (with the number n of choices) than the T completely independent values. (This is true whether we consider them as belonging to the set $\mathbb{Z}_n$ of $m = n$ elements or to the subset $\Gamma(n)$ of $m = \varphi(n)$ elements, since m grows "on the average" as cn.)

To be precise, we can formulate this observation as a hint for the presence of some particular *repulsion* of the geometric progressions $\{2^t \pmod{n}\}$ from each other (which has not yet been investigated). Because of this repulsion an orbit is far from random, and coincidences of the points $2^t \pmod{n}$ with each other (or even close approaches) turn out to be fewer than if the sequence of points ($t = 1, 2, \ldots, T$) were chosen independently of each other in the m-element set ($\mathbb{Z}_n$ or $\Gamma(n)$).

9. Measurement of the Degree of Randomness of a Subset

I sought to measure the degree of randomness of the points of an orbit $\{a^t\}$ of the Fermat-Euler transformation among all the residues from $\mathbb{Z}_n$ (or among the $m = \varphi(n)$ residues relatively prime to n which make up the Euler group $\Gamma(n)$). I decided to measure the "repulsion" from each other of the elements of a T-element set of points on the segment using the following quantity.

Denote by $\{x_1, \ldots x_T\}$ the sequence of distances between consecutive points.

Having in mind an application to residues, I glued the ends of the segment into a circle, so that the sum of the positive numbers x_i is equal to the length L of this segment or circle.

To measure the presence or absence of near approaches of points of the set I used a *"parameter of randomness"*

$$R = x_1^2 + \cdots + x_T^2.$$

In order to make this parameter dimensionless (independent of units of measurement; that is, of the length L), I normalized it, as if transforming it to the case $L = 1$. The *normalized parameter of randomness T of points is*

$$r = r/L^2.$$

This normalization is necessary in order to apply the theory to the residues of $\mathbb{Z}_n$ (where $L = n$) or to the elements of the Euler group $\Gamma(n)$.

(Here $L = m$, and the "distances" x_i are the numbers of arcs free of the elements of the group $\Gamma(n)$ between two elements of this group, the distance between which is measured by the integer x_i).

All possible configurations of T points of a circle of length 1 are described by the $(T-1)$-dimensional simplex

$$\{x = (x_1, \dots, x_T),\ 0 \leq x_8 \leq 1,\ \sum x_i = 1\}$$

(up to rotations of the circle).

The parameter of randomness r is the moment of inertia of point x with respect to the origin (the square of the distance from x to 0). Therefore its smallest value corresponds to the center of the simplex: $x_i = 1/T$, $r_{\text{min}} = T(1/T)^2 = 1/T$, and its greatest value is at the vertex ($x_1 = 1$, the others are zeroes); that is, $r_{\text{max}} = 1$.

To compare the set with various numbers of elements of T it is natural to introduce the *binormalized parameter of randomness*

$$s = r/r_{\text{min}} = T\sum(x_i/L)^2.$$

The minimal value $s = 1$ is achieved at a regular barracks-style arrangement (an arithmetic progression) of the vertices of a regular T-gon $x_i = 1/T$ (if we consider $L = 1$). In this case we can speak of a "strong repulsion", which won't allow points to get near each other.

The greatest value $s = T$ is achieved at a dense clustering of T points that coincide. Here the distances x_i are equal to zero, except for one distance, which is equal to one. In this case we can speak of a "strong attraction", drawing all the points to one location.

We now look at some truly random distributions of T independently distributed points on the circle. It turns out that the binormalized parameter of randomness of such distributions has a completely determined value s_1 which is close to $s = 2$ for large numbers T of points.

We now compute this value s_1, an "indicator of a chaotic set". We can call it a "freedom-loving value", which suggests an absence of either repulsion or attraction to their neighbors of points of the set considered. Smaller values of the parameter, right down to $s_{\text{min}} = 1$, correspond to a lattice-like regular structure of equidistant points, indicating some repulsion.

Larger values of the binormalized parameter of randomness than the freedom-loving value $s \approx 2$ indicate a mutual attraction of the points of the set. An extreme example of this phenomenon is the clustering that corresponds to the maximal value of T (for a fixed number of elements of the set).

10. The Average Value of the Parameter of Randomness

Consider a random distribution of points x in the simplex of dimension $T-1$

$$\{0 \leq x_i \leq 1, \sum_{i=1}^{T} x_i = 1\},$$

of constant density with respect to the Lebesgue measure density (which also corresponds to the distance between T points, independently and randomly scattered on the circle of length 1).

Theorem. *The average value of the parameter of randomness* $s = T\sum_{i=1}^{T} x_i^2$ *is equal to the "freedom-loving value"*

$$s_1 = \frac{2T}{T+1}.$$

Proof. We compute the average value for each addend x_i^2 and add these averages (using their independence).

The volume of the layer of our simplex between $x_i = u$ and $x_i = u+\varepsilon$ is equal, in a first approximation with respect to ε, to the product $C\epsilon(1-u)^{T-2}$ where C is the volume of the unit $(T-2)$-dimensional simplex (because this $(T-1)$-dimension layer of thickness ε rests on the $(T-2)$-dimension simplex $\{0 \leq x_j \leq 1-u, \sum x_j = 1-u$, where $j \neq i\}$, which has volume $C(1-u)^{T-2}$).

Hence the integral of x_i^2 over all of the $(T-1)$-dimension simplex is

$$I = \int x_i^2 dx^{T-1} = \int_{u=0}^{u=1} Cu^2(1-u)^{T-2}du = \int_0^1 CV^{T-2}(1-v)^2dv.$$

This last integral is easy to compute (with a necessary transition to the self-similar variable v) and is equal to $I = C\left(\frac{1}{T-1} - \frac{2}{T} + \frac{1}{T+1}\right)$. Also, the volume of our whole $(T-1)$-dimension simplex is equal to an analogous integral without the factor u^2; that is, without $(1-v)^2$:

$$M = \frac{C}{T-1}.$$

Thus, the average value of the sum T of the addends TX_i^2 is $s_1 = T^2I/M = T^2\left(1 - \frac{2(T-1)}{T} + \frac{T-1}{T+1}\right) = \frac{2T}{T+1}$, which proves the theorem.

Comparing the observed values of the parameter s of randomness for geometric sequences with the freedom-loving value s_1, I found values smaller than the freedom-loving value, $1.4 \leq s \leq s_1$ in the majority of progressions $\{2^t \pmod{n}\}$. Usually, these values are close to 1.6. This suggests a noticeable repulsion between the residues of the elements of the geometric progressions.

But I have proven no theorem about the asymptotic value of $s(n)$ for large n.

Remark. It would be interesting to look not only at the average value s_1, but also at other characteristics of truly random sets. Some examples are the distribution function of the value x_i, or its moments, or the distribution of probabilities of various partitions of the sum L into T integers x_i (in the case of integer values of the variable x_i).[14]

All these integrals can be computed explicitly and give piecewise-polynomial distributions (for example, for the quantity s, which depends on a random parameter x running through a simplex).

It is likely that these distributions (and especially their universal asymptotics near the freedom-loving average value s_1 for large T) are worthy of special study in probability theory and stochastic geometry.

As for the statistics of the residues of geometric progressions, the distributions here turn out to be different, and their study, even experimentally, might bring new discoveries in ergodic number theory.

Aside from geometric progressions, I have also tried to study arithmetic progressions of residues $\{dt \pmod n),\ t = 1, 2, 3, \ldots, T\}$, and the set of residues relatively prime to n, $\Gamma(n) \subset \mathbb{Z}_n$.

In both cases the observed value of the parameter of randomness lies between 1 and 2, which may be regarded as an evidence of a repulsion of neighbors from each other. It is likely, in the case of arithmetic progressions, that these results can be proven using continued fractions (and underlying generalizations of Kuzmin's theorem).

11. Additional Remarks about Fermat-Euler Dynamics

An experimental investigation of the functions T and N of the variable n led me to hundreds of observations, some of which have now become theorems. Here is the simplest example.

Definition. The odd number n is said to belong to the class $(N+)$ if the following Fermat-Euler equation is satisfied:

$$2^{\varphi(n)/N} \equiv 1 \pmod n.$$

Theorem. *The class $(N+)$ is an ideal in the commutative multiplicative semigroup of odd numbers: if n belongs to the class $(N+)$, then the product of n and any odd natural number also belongs to the class $(N+)$.*

---

[14] I mention here that it was proven in [**16**] that the probabilities p_k of finding arcs of length k among T arcs into which T distinct, randomly chosen points divide the finite circle $\mathbb{Z}_m$ are proportional to the binomial coefficients appearing in lines parallel to the sides of Pascal's triangle, at a distance $T-2$ from its edges:

$$p_k = \binom{m-1-k}{T-2} \Big/ \binom{m-1}{T-1} \quad (1 \le k \le m-T-1).$$

For example, for $T = 4$ the probabilities of the lengths of the arcs for random points are to each other as

$$1 : 3 : 6 : 10 : 15 \cdots .$$

(The largest arc length, which is $m - T - 1$, has the smallest probability.)

Example. The numbers **31**, **43**, 63, 91, 93, 117, 129, 133, 155, **157**, 171, 189, 215, 217, **223**, **229**, 247, 259, 273, 279, **283**, 301 belong to the class (3+). (Prime numbers are in boldface.)

The generators of the semigroup are those numbers which are not divisible by any others; that is, all the primes and also 63, 91, 117, 133, 171, 247, 259.

A strange observation, for which no explanation has yet been found, is that the residues modulo 9 of all these generators are quadratic residues. (They belong to the set $\{0, 1, 4, 7\}$).

An analogous result holds for the class (5+) and residues modulo 25. (Just as with the previous comment, this is only an observation from tables we have, although these tables reach quite far.)

For several other values of N, not every generator of the ideal $(N+)$ is a quadratic residues modulo N^2, but only those that are prime.

The class $(N-)$ is defined by the condition

$$2^{\varphi(n)/N} \equiv -1 \pmod n$$

on its elements n.

If the number $\varphi(n)$ is divisible by 4 and the odd number n belongs to the class (2+), then n belongs either to (4+) or to (4−), because

$$(2^{\varphi(n)/4} - 1)(2^{\varphi(n)/4} + 1) = 2^{\varphi(n)/2} - 1 \equiv 0 \pmod n$$

by Euler's theorem.

But we cannot say explicitly which of the elements of (2+) belong to (4+), and which to (4−), just as we cannot distinguish the subclasses (8+) and (8−) within the class (4+).

There are sometimes quite explicit conditions for a number to belong to various types of classes. For example, the following theorem is proved in [**16**]:

Theorem *If the odd number n has k or more different prime divisors, then it belongs to the class* (2^k+).

Sometimes significant information about the class of a number or a product is given by a description of the classes of its factors. For example, in [**16**] and [**12**] we have the theorem (probably going back to Euler, if not to Fermat):

Theorem. *The odd number p^a is in the class* (2+) *if the prime number p gives a remainder of* 1 *or* −1 *upon division by* 8, *and in the class* (2−) *if it gives a remainder of* 3 *or* −3.

Hence it is not so hard to find whether an odd number belongs to the class (2+) or (2−). But even for the classes (4+) and (4−) (and even for the primes in these classes) the situation is more complicated and the criteria are less clear.

For example, the numbers

$$\mathbf{17}, \mathbf{41}, 57, \mathbf{97}$$

belong to the class (4−), while the numbers

$$65, \mathbf{73}, \mathbf{89}, \mathbf{113}$$

belong to the class (4+). But the reason for this is unclear.

The tables hint at dozens of different conjectures, some of which have already become theorems.

For example, it is proven in [**16**] that *if a prime number* $p = 8c+1$ *(such as* $p = 73$*) belongs to the class* (4+)*, then every product* $p^a q^b$*, where* q *is another odd prime, belongs to the class* (8+).

12. Primitive Roots of a Prime Modulus

Here we will prove Theorem 2 from Section 4, of which we have proved only half so far: we have proved that $a^{p-1} \equiv 1 \pmod{p}$ for a prime p.

It remains to prove that the group $\Gamma(p)$ is cyclic; that is, that there exists a geometric progression $\{a^t\}$ mod p, such that all its $p-1$ elements (for $1 \le t \le p-1$) are distinct (so that $a^t \not\equiv 1 \pmod{p}$ for $0 < t < p-1$).

Such a generator $a \in \Gamma(p)$ is called a *primitive root.* It turns out that the number of such roots is $\varphi(p-1)$.

The proof of the existence of a primitive root is based on a remarkable identity of Euler: *the sum of the values of the Euler function over all the divisors* d *of the number* n *is equal to* n *itself:*

$$\varphi(1) + \cdots + \varphi(d) + \cdots + \varphi(n) = n.$$

For example, the divisors of 6 are $d = 1, 2, 3, 6$, and the Euler identity becomes:

$$\varphi(1) + \varphi(2) + \varphi(3) + \varphi(6) = 1 + 1 + 2 + 2 = 6.$$

We get a proof of Euler's identity from the decomposition of all the residues modulo n depending on their greatest common divisors with n.

Residues modulo n whose greatest common divisor with n is d have the form kd, where the number k must be relatively prime to n/d (otherwise the common divisor d of the residue with n would not be the greatest).

Therefore the total number of residues can be represented in the form of a sum over all the divisors d of n:

$$n = \sum_d \varphi(n/d).$$

But since the number $k = n/d$ is also a divisor of n, and complementary divisors d and k determine each other, the last sum can be rewritten in the form

$$n = \sum_k \varphi(k),$$

where the sum is again taken over all divisors of n. This proves Euler's identity.

Example. For $n = 6$ the residues which have the greatest common divisor $d = \{1, 2, 3, 6\}$ with $n = 6$ are $(1, 5)$, $(2, 4)$, (3), (6) respectively. The number of residues having these greatest common divisors with $n = 6$ is equal to $\varphi(6/1) = 2$, $\varphi(6/2) = 2$, $\varphi(6/3) = 1$, $\varphi(6/6) = 1$, respectively, so that the total number of residues ($n = 6$) is represented by the partition described into classes with different greatest common divisors of d and n in the form:

$$6 = \varphi(6/1) + \varphi(6/2) + \varphi(6/3) + \varphi(6/6) = \varphi(6) + \varphi(3) + \varphi(2) + \varphi(1).$$

Now let us consider geometric progressions of the form $\{a^t \mod p\}$, where p is an odd prime and a is relatively prime to p. By Fermat's Little Theorem, we have $a^{p-1} \equiv 1 \pmod{p}$. But for some bases a it might happen that the minimal period of the progression is not $p-1$ but rather some divisor T of the number $p-1$:

$$a^T \equiv 1 \pmod{p}.$$

In this progression there are $\varphi(T)$ elements $b = a^t$ (where $0 < t < T$ is relatively prime to T) for which for the progression $\{b^t\} \pmod{p}$. has the same minimal period T

Indeed, $b^T = a^{tT} = (a^T)^t \equiv 1 \pmod{p}$, and there can be no smaller period than T, since $b^r = a^{tr} = a^u$, where u is the remainder when tr is divided by T, which is less that T. Thus if b^r were congruent to 1 modulo p, then the period T would not be minimal for the base a either.

There are no solutions to the congruence $a^T \equiv 1 \pmod{p}$ other than the T elements of the progressions $\{a^t\}$, $\pmod{p}$, because a congruence of degree T modulo a prime p and leading coefficient 1 cannot have more than T solutions (by Vieta's theorem). This means that $\varphi(T)$ is the total number of solutions of this congruence.

Thus the distribution of all $p-1$ numbers $0 < a < p$ (which are all relatively prime to p) according to the minimal period T of the progressions $\{a^t \pmod{p}\}$ (which are all divisors of the number $p-1$) has the form

$$p - 1 = \Sigma\varphi(T), \qquad (*)$$

where the sum is taken over those divisors T of $p-1$ for which T is the smallest period of one of the progressions $\{a^t\}$.

By Euler's identity, the number $n = p-1$ is equal to the sum of the values of $\varphi(d)$ over *all* the divisors d of $p-1$. This means that no divisor d is left out of the sum $(*)$. This concludes the proof of the following theorem:

Theorem. *For any divisor d of $p-1$, the number of residues $0 < a < p$ with minimal period d for the progression $\{a^t \pmod{p}\}$ is equal to $\varphi(d)$.*

In particular, the number $d = p-1$ is such a divisor.

Corollary. *The number of primitive roots modulo p is equal to $\varphi(p-1)$ (and, in particular, such roots always exist).*

Example. For the modulus $p = 7$ the number of primitive roots is $\varphi(6) = 2$. The progressions $\{a^t \pmod{7}\}$ $(t = 1, 2, \ldots)$ (where $a = 1, 2, \ldots, 6$)

and their smallest periods $T(a)$ are:

$$\begin{aligned}&\{1, 1, \dots\},\ T = 1;\\&\{2, 4, (8 \equiv 1); 2, \dots\},\ T = 3;\\&\{3, (9 \equiv 2), 6, (18 \equiv 4), (12 \equiv 5), (15 \equiv 1); 3, \dots\},\ T = 6;\\&\{4, (16 \equiv 2), (8 \equiv 1); 4, \dots\};\ T = 3;\\&\{5, (25 \equiv 4), (20 \equiv 6), (30 \equiv 2), (10 \equiv 3,)(15 \equiv 1); 5, \dots\},\ T = 6;\\&\{6, (36 \equiv 1); 6, \dots\},\ T = 2.\end{aligned}$$

The primitive roots are $a = 3$ and $a = 5$. The number of roots with period $T = 3$ is equal to $\varphi(3) = 2$.

13. Patterns in Coordinates of Quadratic Residues

We now use the facts proven above about geometric progressions to describe the geometry of quadratic residues for odd prime moduli p.

Let us denote by A some primitive root mod P, $0 < A < p$. The geometric progression $\{A^t\}$, $0 < t < p-1$, contains all the non-zero residues modulo p, once each. The non-zero quadratic residues correspond to even values of t.

We denote by T the smallest period of the geometric progression $\{2^t\}$, $2^T \equiv 1 \pmod p$, and we denote by N the number of rows in the Young diagram of the permutation "multiplication of residues by 2" of the set of all $\varphi(p) = p - 1$ non-zero residues modulo p (so that $TN = p - 1$).

Basic Proposition. *All TN residues* $\pmod p$ *of the numbers*

$$A^T 2^n \quad (0 \le s < T,\ 0 \le r < N)$$

are distinct.

Proof. If two residues were to coincide, then one of our residues (namely, the quotient of the two residues which had coincided) would be equal to one:

$$A^u 2^v \equiv 1 \pmod p, \quad (0 \le u < t,\ 0 \le v < N).$$

If u and v are not both zero, then raising this congruence to the power T, we would have the congruence

$$A^{Tu} 2^{Tv} \equiv A^{Tu} \equiv 1 \pmod p, \quad \text{where } 0 < Tu < TN = \varphi(p),$$

That is, the residue A would not be primitive for modulo p.

Thus we have proved our Basic Proposition. We will use the residues $s \pmod T$ and $r \pmod N$ as *coordinates* of points in a Young diagram, or of the residues $A^t \equiv A^r 2^s \pmod p$.

Let us describe *the places occupied by the quadratic residues* using these coordinates. They form a beautiful pattern on the (r, s)-plane. By their very definition, all residues with even r and s are quadratic residues. But the number of such residues is only about a quarter of the whole number

$p-1$ of non-zero residues, while the number of quadratic residues is equal to half of that;[15] that is, to $(p-1)/2^3$.

This means that there must be still more quadratic residues, arising from the squares of the residues $a^i 2^j$. These are equal to $A^{2i}2^{2j} \equiv A^r 2^s \pmod{p}$, where $r \neq 2i$.

The squares with coordinates (r, s) are situated on the Young diagram in different ways, depending on the remainder when the prime modulus p is divided by eight.

Theorem. *If $p = 8c \pm 1$, then all the residues on even rows $A^{2r}2^s$, and only those, are quadratic residues. Furthermore, in this case the number of rows N is always even.*

If $p = 8c \pm 3$, then the quadratic residues occupy half of each row; namely, the residues $A^{2r}2^{2s}$, $A^{2m-1}2^{2m-1}$, and these are the only quadratic residues. Furthermore, in this case the number of rows N is always odd, and the length T of each row is even (it is a multiple of 4 for $p = 8c-3$ and is not a multiple of 4 for $p = 8c+3$).

A few examples of values of the number of rows N and their lengths T are shown in the upper set of tables below.

The interesting question of the growth of the numbers T and N as p increases has been studied only empirically. Experiments indicate that the values of N are often much less than those of T (the values of N may even remain bounded, on the average).

We will consider four examples (with the number p having remainders 1, 3, 5, 7 modulo 8).

$p = 8c + 1$:

c	p	T	N
2	17	8	2
9	73	9	8
14	113	28	4
29	233	29	8

; $p = 8c + 3$:

c	p	T	N
0	3	2	1
1	11	10	1
5	43	14	3
31	251	50	5

;

The quadratic residues modulo p are printed in boldface in the following four Young diagrams. In each diagram, the rows show the residues of the numbers $A^r 2^s$ for a fixed r. (Here the coordinate $s = 0, 1, \dots, T-1$ increases from left to right, while the coordinate $r = 0, 1, \dots, N-1$ increases from top to bottom, as is usual for matrices.) Looking at these tables, it is easy to see that the number $A = 3$ is a primitive root modulo $p = 17, 43$, and 31.

The previous theorem asserts that the pattern formed by the quadratic residues in these diagrams is not an accident: the proof given below shows

[15] The operation of squaring a residue folds the set of non-zero residues in half because the number 1 is the square only of 1 and -1, and consequently every non-zero quadratic residue is the square of two different residues, $\pm x$.

$p = 8c + 5$:

c	p	T	N
0	5	4	1
1	13	12	1
4	37	36	1
13	109	36	3

; $p = 8c + 7$:

c	p	T	N
0	7	3	2
3	31	5	6
15	127	7	18
53	431	43	10

.

$p = 17: \ N = 2, T = 8,$

1	**2**	**4**	**8**	**16**	**15**	**13**	**9**	$(18 \equiv 1)$
3	6	12	7	14	11	5	10	$(20 \equiv 3)$

;

$p = 43: \ N = 3, T = 14,$

1	2	**4**	8	**16**	32	**21**	42	**41**	39	**35**	27	**11**	22	$(44 \equiv 1)$
3	**6**	12	**24**	5	**10**	20	**40**	37	**31**	19	**38**	33	**23**	$(46 \equiv 3)$
9	18	**36**	29	**15**	30	**17**	34	**25**	7	**14**	28	**13**	26	$(52 \equiv 9)$

;

$p = 13: \ N = 1, T = 12,$

1	2	**4**	8	**3**	6	**12**	11	**9**	5	**10**	7	$(14 \equiv 1)$

;

$p = 31: \ N = 6, T = 5,$

1	**2**	**4**	**8**	**16**	$(32 \equiv 1)$
3	6	12	24	17	$(34 \equiv 3)$
9	**18**	**5**	**10**	**20**	$(40 \equiv 9)$
27	23	15	30	29	$(58 \equiv 27)$
19	**7**	**14**	**28**	**25**	$(50 \equiv 19)$
26	21	11	22	13	(26)

.

that exactly this distribution of quadratic residues (which varies depending on the remainder upon division of the modulus p by 8) is inevitable.

Proof of Theorem. Notice first that if the number N is even, then there are no quadratic residues in the line $A^r 2^s$ for odd values of r. Therefore in the rows with even r there must be quadratic residues aside from those for which s is even (which are obviously quadratic residues).

If any residue $A^{2r}2^{2n-1}$ is quadratic, then all the other residues with exponents of the same parity,

$$A^{2u}2^{2v-1} = A^{2r}A^{2n-1}(A^{u-r}2^{v-n})^2,$$

are also quadratic.

Thus *for even N the quadratic residues are all the residues in the even rows, $\{A^{2n}2^s\}$, and only those.*

But if the number N of rows is odd, then, as we will soon prove, *the quadratic residues form half of each row; these are the residues $\{A^r2^s\}$, where r and s have the same parity.*

To prove this, we denote the odd number N by $2r-1$. (We note that the period T is even for odd N, since the product $NT = p-1$ is even.)

The square of the element A^r is $A^{N+1} = A^N A$.

Lemma 1. *The following relation holds for the elements of the Nth and zero rows:*

$$A^N \equiv 2^i \pmod p,$$

where i is some integer, $0 < i < T$.

Proof. The products of the form

$$A^u 2^v \quad (\text{where } 0 \le u < N, 0 \le v < T)$$

exhaust the $NT = \varphi(p)$ residues modulo p, according to the Basic Proposition proved above. Therefore the residue A^{N+1} must coincide with one of these.

By the same Basic Proposition, it cannot coincide with residues of an element $A^w 2^i$ of some intermediate row (for which $0 < w < N$). Therefore $w = 0$ and the lemma is proved.

Now we can represent the residue of the square of some element A^r in the form $A2^i$, which is in the first row. We conclude that all the residues of elements in each row $A^u 2^{iu}$ are also squares, and therefore so are all the elements of the form $A^u 2^j$, where j has the same parity as iu.

Thus we have obtained $T/2$ quadratic residues in each of the N rows; that is, $\varphi(p)/2$ non-zero quadratic residues in all. This means we have obtained all the quadratic residues.

In addition, we can conclude that the number i is odd, since otherwise the residue A itself would have to be quadratic, so that $A \equiv A^{2s} \pmod p$, $A^{2s-1} \equiv 1 \pmod p$, and the odd number $2s-1$ would have to be divisible by the even period $\varphi(p)$ of the operation of multiplication of residues by the primitive root A.

For an odd number N of rows, the fact that i is odd implies that the exponents u, v have the same parity for the quadratic residue $A^u 2^v$.

To conclude the proof of the theorem, we study how the parity of the number N of rows depends on the remainder when the modulus p is divided by 8.

Lemma 2. *If $p = 8c \pm 3$, then the number N of rows is odd.*

Proof. If the number N of rows were even, then for the prime p as in Lemma 2 we would find, respectively,

$$\begin{array}{cc} \varphi = 8c+2 = 2(4c+1); & \varphi = 8c-4 = 4(2c-1); \\ N = 2m; & N = 2m \text{ or } N = 4m; \\ 4c+1 = mk; & 2c-1 = mk; \\ T = k & T = 2k \text{ or } T = k \end{array}$$

(where the number k is always odd, since it is a divisor of the odd number $\varphi/2$ or $\varphi/4$ respectively, equal to mk).

From these formulas we obtain the relation

$$2^{\varphi/2} = 2^{mk} \equiv (2^T)^m \equiv +1 \quad \text{for} \quad p \equiv 3 \pmod 8.$$

But if $p = 8c-3$, then we have either the congruence

$$2^{\varphi/2} = 2^{2mk} \equiv (2^T)^m \equiv +1$$

in the first case indicated above (when $N = 2m$), or the congruence

$$2^{\varphi/2} = 2^{mk} \equiv (2^T)^2 m \equiv +1$$

in the second case, when $N = 4m$. In all three cases, this contradicts the property $p \in (2-)$ of the prime numbers $p = 8c\pm 3$; that is, the Fermat-Euler congruence

$$2^{\varphi(p)/2} = \quad 1 \pmod p.$$

which they satisfy (see, for example, [**12**]). This proves Lemma 2.

Lemma 3. *If $p = 8c \pm 1$, then the number N is even.*

Proof. Let $p = 8c-1$. Then $\varphi = 8c-2 = 2(4c-1)$. Then if N were odd, we would find that the period would be even: $T = 2m$. From this we have $4c-1 = mk$, and $N = k$.

Then, for the sequence of residues $\{2^i \pmod p\}$ we would have a period $T = 2m$ (by the theorem of Euler-Fermat, which says that $p \in (2+)$), and $\varphi(p)/2 = mk$). From the odd parity of this last number it follows that the period T is not minimal, which contradicts its definition. This means that the assumption that there are oddly many rows N is false, and the lemma is proved for the case $p = 8c-1$.

The proof for the case $p = 8c+1$ is more complicated, and we will first examine several auxiliary constructions.

From the theorem of Fermat-Euler, we have the following congruence (proved, for example, in [**12**]):

$$2^{\varphi(p)} - 1 \equiv 0 \pmod p.$$

Let us represent the number $\varphi(p) = 8c$ in the form of a product $\varphi(p) = 2^a n$, where the number n is odd (and $a \geq 3$). Factoring the differences of two squares we can rewrite the Fermat-Euler congruence as

$$(2^{t_1}+1)\cdots(2^{t_i}+1)\cdots(2^n+1)(2^n-1) \equiv 0 \pmod p,$$

where $t_i = 2^{a-i}n, \ 1 \leq i \leq a$.

One of these expressions in parentheses is the zero residue, and we can distinguish two cases.

Case I. The congruence

$$2^n \equiv 1 \pmod p$$

holds.

Assertion. *In this case the period T is an odd divisor of the number $n = Tm$, while the number of rows is even: $N = 2^a m$.*

Indeed, by condition I the odd number n is one of the periods of the operation of multiplying the residues by 2. This means that it is divisible by the smallest period T, which is therefore odd. This means that the number N of rows is even, since $TN = \varphi(p)$ is an even number.

Case II. The congruence

$$2^{t_i} \equiv -1 \pmod p.$$

Assertion. *In this case the number N is even.*

Indeed, squaring congruence II, we see that the number $2t_i = 2^{a-i+1}$ is one of the periods of the operation of multiplying the residues by 2 and therefore is divisible by the smallest period T.

On the other hand, the minus sign in congruence II shows that the number $t_i = 2^{a-i}n$ itself is not divisible by the period T. This means that the number T is divisible by 2^{a-1+1} and therefore is of the form

$$T = 2^{a-i+1}m,$$

where m is an odd divisor of the odd number $n = mk$.

Thus the number $N = \varphi(p)/T = 2^a mk/(2^{a-i+1}m) = 2^{i-1}k$ must be even, if $i > 1$.

If the number i were equal to 1, then we would have

$$2^{\varphi(p)/2} \equiv -1 \pmod p,$$

in contradiction to the Euler-Fermat theorem $(8c + 1 \in (2+))$, according to which (see, for example, article [2])

$$2^{\varphi(p)/2} \equiv +1 \pmod p.$$

Thus we have proven that N is even, which concludes the proof of Lemma 3.

The information we now have about the parity of the numbers T and N, together with the proof of the initial theorem analyzing the dependence of the coordinates u and v of quadratic residues $A^u 2^v$ on the parity of T and N, concludes the proof of the theorem.

14. Applications to Quadratic Congruences

Wonderful results about the representation of numbers by quadratic forms follow immediately from the description of the pattern of coordinates of quadratic residues proved in the previous section. (A description of this theory and its connection to the relativistic world of De Sitter is discussed in article [**13**].)

Theorem 1. *If the number x^2+y^2 is divisible by the prime number $p=4c+3$, then x and y are also divisible by p.*

In other words:

Theorem 1′. *The congruence $x^2+y^2\equiv 0 \pmod p$ has no non-zero solutions $x \pmod p$, $y \pmod p$, if p has remainder 3 upon division by 4.*

Corollary. *If the equation $x^2+y^2=n$ has integer solutions, and $n=\prod p_i^{a_i}\prod q_j^{b_j}$ is the prime factorization of the right hand side n, where $p_i\equiv 3 \pmod 4$ for every i, then there also exists an integer solution for any equation without factors p_i in the right hand side; that is, for*

$$x^2+y^2=m,$$

where $m=\prod q_j^{b_j}$.

Remark 1. In particular, none of the numbers $n=3,27,21,63$ can be represented in the form x^2+y^2 for integers x and y.

Remark 2. In fact all prime numbers $q\equiv 1 \pmod 4$ (and consequently, according to the result in [**13**], all products of their powers) can be represented in the form x^2+y^2.

The solvability of such equations for non-zero residues follows from the results of the previous section, but we will not prove statements about the actual representation $q=x^2+y^2$ (for example, $5=4+1$, $13=9+4$, $17=16+1$). We limit ourselves to the investigation of congruences.

Theorem 2. *If the number x^2+2y^2 is divisible by the prime number $p=8c+5$ or $8c+7$, then the integers x and y are also divisible by p.*

As in the case of Theorem 1′ and its corollary, Theorem 2 reduces the solution of the equation $x^2+2y^2=n$ to the case where each prime factor p of the number n gives a remainder of 1 or 3 upon division by 8. In this case the equation $x^2+2y^2=p$ in fact has integer solutions (from which it follows, according to the result in [9], that equations with any right hand side n, divisible only by such prime numbers also have solutions.) We will not prove this, but limit ourselves to congruences whose solutions follow easily from the previous section (Theorem 2).

Example. For $p=5$, substituting all possible values of the residues x,y into the quadratic form $x^2+y^2 \pmod 5$ give us a matrix.

The sum x^2+2y^2 is zero modulo 5 in only one case, when $x\equiv 0\equiv y$ $\pmod 5$.

Theorem 2 generalizes this observation to the case of any prime number p in place of 5, but with the condition that it give a remainder of 5 or 7 when divided by 8, in which case the matrix will include only one zero.

For $p\equiv 1$ or $3 \pmod 8$, the situation is completely different: a solution exists. (For congruences, rather than equations, we can get a full solution

$y \backslash x$	0	1	2	3	4
0	0	1	4	4	1
1	2	3	1	1	3
2	3	4	2	2	4
3	3	4	3	3	4
4	2	3	1	1	3

from the results of the previous section.) Jacobi, Euler, and Fermat all studied this equation.

Example. The quadratic form x^2+2y^2 can represent the prime numbers

$$17 = 3^2 + 2 \cdot 2^2 \equiv 1 \pmod 8,$$

$$19 = 1^2 + 2 \cdot 3^2 \equiv 3 \pmod 8,$$

and in general *all prime numbers congruent to 1 or 3 modulo 8, and also all their products.* (The assertion about the products is proved in [**13**].)

It has been proven that the question of the representation of integers by a quadratic form reduces to a question about congruences: *if an equation of degree two has a solution as a congruence for sufficiently many moduli, then it has an actual integer solution* ("Hasse's principle").

For Diophantine equations of higher degrees we find, on the contrary, cases where the congruence is solvable for any modulus at all, but there is no actual solution in integers. (I don't know how often this occurs.)

The question here is similar to the problem of the convergence of formal power series for the solution of problems in analysis. The existence of a formal series solution for an equation implies the existence of a solution modulo arbitrarily high degrees of the variables. But in general, the existence of an analytic solution does not follow: the series might diverge.

Proof of Theorem 1. First we suppose $p = 8c + 3$. We use the description of the pattern of coordinates of remainders upon division by p of the non-zero quadratic residues

$$x^2 = A^{2r}2^{2s} \quad \text{or } A^{2r-1}2^{2s-1} \quad \text{for } p = 8c+3.$$

By the theorem of Euler-Fermat we have $p \in (2-)$; that is, the congruence

$$(2^{\varphi/2} = 2^{4c+1}) \equiv -1 \pmod p$$

holds.

From this congruence we conclude that the residues opposite the squares form an additional pattern,

$$-y^2 \equiv A^{2r-1}2^{2s} \quad \text{or } A^{2r}2^{2s-1}.$$

Therefore the congruence $x^2+y^2 \equiv 0$, that is, $x^2 \equiv -y^2 \pmod{p}$, has no non-zero solutions, and *Theorem 1 is proved for the case* $p \equiv 3 \pmod{8}$.

For $p = 8c+7$, the non-zero quadratic residues have the form $x^2 \equiv A^{2r}2^s$ (by the theorem on the coordinates of quadratic residues).

From the congruence $x^2+y^2 \equiv 0 \pmod{p}$ for $y^2 = A^{2u}2^v$ we would have

$$A^{2u}2^v(1 + A^{2(r-u)}2^{s-v}) \equiv 0 \pmod{p},$$

which, as we shall soon see, is not possible.

Indeed, only one power of a primitive root A can have the residue -1 modulo p; namely, $A^{\varphi/2} = A^{4c+3}$ (because the square of this quantity must have residue 1, so that twice the exponent must be divisible by the period $\varphi(p)$ of the progressions $\{A^t\}$). Thus, for $x^2+y^2 \equiv 0 \pmod{p}$, we would have the congruence

$$A^{2(r-u)}2^{s-v} \equiv A^{4c+3} \pmod{p}.$$

Using the congruence $2^T \equiv 1 \pmod{p}$ and the definition of the number N of rows leads us to a congruence of the form

$$A^d 2^3 \equiv 1 \quad (0 \le d < n, 0 \le e < T)$$

with an odd exponent d. This last congruence contradicts the Basic Proposition of the previous section, which tells us that all NT residues of this form are distinct.

Thus it must be true that *the congruence* $x^2+y^2 \equiv 0 \pmod{p}$ *has no non-zero solutions for* $p = 8c+7$

This proves Theorem 1. (Whether Fermat discovered it first, or Euler later, I have not been able to figure out.)

Proof of Theorem 2. Suppose again $p = 8c+7$, so that the pattern of squares has the form

$$x^2 \equiv a^{2r}2^s, \quad y^2 \equiv A^{2u}2^v \pmod{p}.$$

In this case we have the congruence

$$x^2 + 2y^2 \equiv A^{2r}2^s + A^{2r}2^{v+l} \pmod{p}.$$

In proving Theorem 1, we have already proved that this congruence has no solutions except when x and y are both simultaneously zero (for the case $p = 8c+7$). This means that *there are no solutions for the congruence* $x^2+2y^2 \equiv 0 \pmod{p}$ *except when x and y are simultaneously zero.*

Next let $p = 8c+5$. In this case, from the theorem about coordinates of quadratic residues, we find the following congruences, which give a pattern for the non-zero squares:

$$x^2 \equiv A^r 2^{2i} \quad (0 \le r < N,\ 0 \le 2i < T),$$
$$y^2 \equiv A^u 2^{2v} \quad (0 \le u < N,\ 0 \le 2v < T).$$

Therefore, from the congruence $x^2+2y^2 \equiv 0 \pmod{p}$ (if x and y are not both zero), we get the congruence

$$A^r 2^{2i} + A^u 2^{2v+1} \equiv 0 \pmod{p}.$$

On the other hand, the theorem of Fermat-Euler "$p \in (2-)$" from [2] implies that the following congruence holds

$$(2^{\varphi(p)/2} = 2^{4c+2}) \equiv -1 \pmod{p},$$

so that we can rewrite the previous congruence in the form

$$A^r 2^{2i+4c+2} \equiv A^u 2^{2v+1} \pmod{p}.$$

This congruence again contradicts the fact that congruences of the form

$$A^d 2^e \equiv 1 \quad (0 \le d < N,\ 0 \le e < T)$$

have no solutions (with $(d,e) \neq (0,0)$). This result is contained in the Basic Proposition of the previous section (which asserts that all NT residues of this form are different).

This proves Theorem 2.

All these examples of applications of the geometry of the patterns formed by quadratic residues $A^d 2^e$ on a plane with coordinates d and e suggest the possibility of applying our results, and their natural generalizations for the progression $\{a^t\}$, to the study of multiplicative semigroups of integers formed by the values taken on by binary quadratic forms $mx^2 + ny^2 + kxy$. For example, if such a quadratic form can represent the number one (as any quadratic form $x^2 + ny^2$ does), then its values form a semigroup. Another example of a semigroup is given by the set of values $\{Nf\}$, where N is one of the values of the form f (say, the values of the form $4x^2 + 2ny^2 = 2(2x^2 + ny^2)$). Many other examples are described in [**13**].

All these multiplicative semigroups would be interesting to study in the geometric terms of Newton's polyhedra, which are formed by vectors of exponents of the degrees which appear in the prime factorizations of the elements of the subgroup by their prime factors:

$$n = 2^a 3^b 5^c 7^d \cdots.$$

The vectors of the infinite dimensional vector space with components $(a, b, c, d, \dots)$, which correspond to the numbers n in the semigroup, certainly form an additive semigroup. It would be interesting to know if it has a description as simple as the description of the semigroup of quadratic forms for the Gaussian quadratic norm $x^2 + y^2$ for the complex numbers . The description of this semigroup is as follows: *the exponents $(b, d, \dots)$ of the primes which give remainder 3 when divided by 4 must be even.*

In the case of the form $x^2 + 2y^2$ the description of the semigroup of values is also simple: *the exponents of the primes which give a remainder of 5 or 7 when divided by 8 must be even.*

But generally speaking, the additive semigroup of integer vectors can have a much more complicated structure (for example, for the semigroup of vectors on a plane or in three dimensional space).

It would be interesting to know if such complicated structures are met with in the semigroups of number theory. Or perhaps the semigroup of values of a quadratic form always admits of a simple description, like the examples above (or like the finite basis of a polynomial ideal).

In both our examples, the restrictions were imposed only on the vector envelope of the additive semigroup (the condition that several coordinates be even), while for a more general additive semigroup restrictions involving inequalities are possible. For example, we can take as our additive semigroup the set of those integer points of some lattice which lie inside a convex cone (or a Newton polyhedron).

I know of no examples of semigroups of values of quadratic forms with such non-trivial Newton polyhedra.

In the actual geometry of additive semigroups, even those consisting simply of natural numbers, there are also some open questions. For example, the semigroup $\{mx + ny\}$, generated by two relatively prime natural numbers x and y (and consisting of linear combinations of these numbers with non-negative integer coefficients), contains all the integers starting with the bound $K = (m-1)(n-1)$ found by Sylvester. But from zero to this bound the semigroup contains exactly half the integers (specifically, if the number c is an element of the semigroup, then $K - 1 - c$ is not an element of it.) Example: if $x = 3, y = 5$, then $K = 8$, and $\{0, 3, 5, 6\}$ are elements of the semigroup, while $\{7, 4, 2, 1\}$ are not.

Generalizations of these results of Sylvester to the case of more than two generating numbers so far remain conjectures. This is true even if we limit ourselves (as is reasonable here) to averaged asymptotics of the quantities like $K(x, y, z)$ for the majority of large vectors (x, y, z), with only rarely encountered larger deviations. (More detail on this is given in [**7**].)

It would also be interesting to study the number of representations of a large element of the semigroup as a sum of the generators. Here too we can probably get beautiful results, if we don't try to pursue exact formulas, but rather limit ourselves to averaged asymptotics, describing the multiplicities of the projections of a set of integral points of the corresponding convex polygon onto typical integer lines (and neglecting large but rare deviations from the mean).

Turning our attention to these averaged asymptotics is the most promising approach to many problems in Diophantine geometry, including problems about integer quadratic forms and integer geometric progressions considered by Fermat and Euler and described above. (See also [**7**].)

This line of investigation can be applied, for example, to studying asymptotics for the smallest period $T(a, n)$ of a geometric progression $\{a^t \pmod{n}, t = 1, 2, \dots\}$ for large values of the modulus n. Is it true that the period $T(2, n)$ grows, on the average, as a power of n? Or in any case faster than $n^{1-\alpha}$ or faster than $n/\ln n$? How does the period $T(3, n)$ grow? What happens to the period as a and n grow simultaneously (for values of a of the order cn)?

Here even simple numerical experiments would be of interest: this is just how Fermat discovered his "Little Theorem", and how Legendre discovered the law of distribution of prime numbers (whose average density is $1/\ln n$).

A table of the periods $T(a, n)$ of geometric progressions $\{a^t (\mathrm{mod}\ n),\ t = 1, 2, \ldots\}$ beginning with the values of the smallest periods is shown below.

In the line labeled S at the bottom of the table, for each modulus n we have given the sum of all the periods: $S = \sum T(a, m)$ for $1 \leq a < m$ (relatively prime to m), for the previous moduli, including n (for $m \leq n$). The summation simplifies the analysis of the asymptotics, playing the role of an average.

In the line labeled M we have given the average period (with respect to a) for each modulus n. We have taken only those values of a which are relatively prime to n.

Tentative conclusions regarding the average behavior indicate that S grows as $cn^2/2$ (with a coefficient c on the order of $1/5$), while the average period M grows as qn (with a coefficient q on the order of $1/3$).

It is interesting to compare these observations with the circumstance that for prime n the maximum value of the period $T(a, n) = \varphi(n)$ is achieved at $\varphi(\varphi(n))$ primitive roots a modulo n. If we consider that the Euler function φ grows on the average as $(6/\pi^2)n$ and if we use (illegally) the previous circumstances for composite n as well, we would find that the contribution of these largest periods to their sum is of the order $6/\pi^2)^3 n$, and the contribution to the sum S of the previous moduli is of the order $(6/\pi^2)^3 n^2/2 \approx n^2/7$.

a																				
20																				
19																			2	
18																		2		
17																	2	9	4	
16																2		9		
15															2	8		18		
14														2		16		18		
13													2	4	4	4	3	18	4	
12												2				16		6		
11											2	12	3	2	4	16	6	3	2	
10										2		6				16		18		
9									2	5		3	3		2	8		9	2	
8								2		10		4		4		8		6		
7							2	3	4	10	2	12		4	2	16	3	3	4	
6						2				10		12				16		9		
5					2	6	2	6		5	2	4	6		4	16	6	9		
4				2		3		3		5		6		2		4		9		
3			2	4		6	2		4	5		3	6		4	16		18	4	
2		2		4		3		6		10		12		4		8		18		
1	1	1	1	1	1	1	1	1	1	1	1	1	1	1	1	1	1	1	1	
	2	3	4	5	6	7	8	9	10	11	12	13	14	15	16	17	18	19	20	n
S	1	4	7	18	21	43	50	71	82	145	152	227	248	271	294	465	486	669	694	
M	1	1.5	1.5	2.75	1.5	3.5	1.75	3.5	2.75	6.3	1.75	6.25	3.5	2.9	2.9	10.7	3.5	10.2	3.1	

As a justification of the illegal transition from prime numbers n to arbitrary numbers, I note that for $\varphi(n)$ a transition from $\varphi(p) = p - 1$ to the average growth $\varphi(n)$ as $(6/\pi^2)n$ will change only the asymptotic coefficient 1 into $6/\pi^2 \approx 2/3$.

For the average period, the same (illegal) reasoning indicates linear growth with the modulus n: the periods of maximal value make up, on the average, something of the order of 2/3 of all the periods $T(a, n)$ (for a fixed modulus n). Thus we can roughly approximate the average period as two thirds of the maximal period, which is $\varphi(n)$. Supposing that this last quantity grows as $(2/3)n$, we get a rough bound on the average of the periods over a on the order of $(4/9)n$. But of course prime values n make up a only small share of all the moduli (and even the value of $\varphi(n)$ for them is of order n, and not $(2/3)n$). Thus to extend the formulas holding for prime moduli to values averaged over all n we must make appropriate modifications (at least in the values of the coefficients). To make these modifications, we would have to extend the table of periods much further.

For geometric progressions with a fixed base a, the table also leads us to empirical data about the average growth of the period $T(a, n)$ as the modulus n (relatively prime to a) increases.

Within the limits of the table given above the approximation $T(2, n) \approx un$, $u \approx 0.38$ looks linear, as it does for the base $a = 3$, where $T(3, n) \approx vn$, $v \approx 0.31$. But this is just an average, and the deviations from this average are rather large ($T(2, 15) = 4$ while $T(2, 19) = 18$). I have made calculations of the period $T(2, n)$ for large moduli n, up to $n = 511$. These indicate an average linear growth as n increases, with a somewhat smaller coefficient u (for large n), and this does not exclude the possibility of the decrease of this "coefficient" as n increases (say, as $1/\ln n$).

For example, $T(2, 511) = 9$, but nearby moduli give periods which are much larger: $T(2, 499) = 166$, $T(2, 503) = 251$, $T(2, 509) = 508$, and the average of the periods $T(2, n)$ over tens of odd moduli, from $n = 493$ to $n = 511$ is approximately 158, which would correspond to a coefficient $u \approx 1/3$.

Computations of values of the periods $T(n)$ and the numbers of orbits $N(n)$ for moduli $n \leq 2001$ have been made by F. Aikardi at my request. These led to the following (amazing) empirical formulas for approximation, which sufficiently describe the growth of these functions on the average:

$$N \sim 0.67n^{2/5}, \quad T \sim 1.41n^{4/5}.$$

These "weak asymptotics" are obtained from linear (non-homogeneous) approximations of the observed dependence of the logarithms of the sums of the values

$$\log\left(\sum_{k-1}^{n} N(k)\right), \quad \log\left(\sum_{k-1}^{n} T(k)\right)$$

on the logarithm of the argument, $\log n$. Here, the sum is taken over odd values of k, if we are studying the period T and the number of orbits N

of a Fermat-Euler system defining a geometric progression of residues $\{2^t \pmod{k}\}$.

Aikardi's tables give, for example, the values for sums shown in the table on the next page. (For $2001 < k \leq 2009$ the data are extrapolated.)

If the number of orbits and the period had asymptotics depending on a power of n,

$$N(n) \approx an^{\alpha}, \quad T(n) \approx bn^{\beta},$$

then for the sums we would get (integral) asymptotics, also depending on a power of n:

$$\sum_{k=1}^{n} N(k) \approx \widetilde{a}n^{\alpha+1}, \quad \sum_{k=1}^{n} T(k) \approx \widetilde{b}n^{\beta+1}.$$

Therefore the empirical data about the behavior of $\sum N$ and $\sum T$ (sums which behave much less chaotically than the wildly oscillating summands N and T themselves) provide an approximation to the values of the coefficients $(a, \alpha; b, \beta)$, which are indicated above.

n	9	109	509	1009	1509	2009
$\sum^{n} N$	5	132	1017	2625	4651	6921
$\sum^{n} T$	15	1409	23607	82761	176016	302277

Remark. One might well be surprised by the values $\alpha \approx 2/5$ and $\beta \approx 4/5$ which we have observed, because the product $\varphi(n) = N(n)T(n)$ grows on the average as cn (where $c = 6/\pi^2 \approx 2/3$).

It might seem that the sum of the exponents of the asymptotics, α and β would then have to be equal to one (the exponent of the weak asymptotics of the product φ)

But the weak asymptotics of the product can differ greatly from the product of the weak asymptotics of the factors, so that the average value of the product can differ greatly from the product of the average values of the factors, especially if large values of the factors are mixed in with small ones.

Example. Suppose that the values of the factor $N(n)$ are mixed up in a neighborhood of the value n of its argument as follows: N takes two values

$$(N_1 = n^u) \gg (N_2 = 1) \ (0 < u < 1),$$

and that the first value appears n^w times more often than the second $(w > 0)$.

To conserve the value of the product $NT = n$ we assume that the second factor must take respectively the values

$$(T_1 = n^{1-u}) \ll (T_2 = n).$$

In this situation the contributions of the indicated neighborhood to the sums $\sum N$ and $\sum T$ are mostly determined respectively by the first part and second part: they are proportional respectively to n^{u+w} and n^1 (if $1-u+w<1$, so that the contribution of $\sum T_1$ is less than the contribution of $\sum T_2$)

Thus the observed empirical weak asymptotics of the factors would be proportional to some power of n, with exponents

$$\alpha = u+w, \quad \beta = 1,$$

whose sum is greater than 1.

For this to happen, we need only the condition noted above that the exponent of dispersion u be larger than the "exponent of mixing" w.

An analysis of the observed values $T(n)$ and $N(n)$ show their significant dispersion. For example, the values of the numbers $N(n)$ change by an order of 10 with a small change in the argument n:

$$N(1960)=2, \quad N(1971)=72, \quad N(1973)=1.$$

A rather good average approximation to the frequencies p_n of values of the quantity $N(n)$ (as the argument changes in the neighborhood of a given value of the modulus n) is given (for $n \leq 2001$) by the following empirical approximate weak asymptotic formulas (where $N=1,2,4,8$, and $N \geq 10$):

$$p_1 \sim C_1 n^{-7/18}, \; p_2 \sim C_2 n^{-1/9}, \; p_4 \sim C_4 n^{1/3}, \; p_8 \sim C_8 n^{1/9}, \; p_{\geq 10} \sim C_{10} n^1.$$

It is clear from these formulas that the ratios of the frequencies of large and small values of N behave (on the average) as a power of n (just as the value of the frequency itself of large and small values does in the example given above).

All these empirical data, which are not supported by any theorems, can be taken as mathematical conjectures. One thing that speaks in their favor is the marvellously exact arrangement of the graphs of the corresponding functions $\sum f$ and $\sum g$, drawn on graph paper with two logarithmic scales ("log-log paper") in the neighborhood of corresponding lines.

We have been talking about the functions (defined by the values $N(k) = y$ and $T(k)=z$, whose occurrence we have been studying):

$$\begin{aligned} f_y(n) &= (k : N(k)=y, \; 1 \leq k \leq n); \\ g_z(n) &= (k : T(k)=z, \; 1 \leq k \leq n). \end{aligned}$$

The linearity, or near-linearity of their graphs on log-log paper indicates the approximate formulas:

$$\log f_y(n) \approx A_y \log n + B_y, \log g_z(n) \approx C_z \log n + D_z,$$
$$(A_y = 1+\alpha_y, \; C_z = 1+\beta_z).$$

The exponents α_y, β_z indicated above for the powers of the asymptotics ($\alpha_1 = -7/18, \dots$) are found simply by drawing these lines, but I know of no *a priori* reason to think that the exponents would be rational (other then perhaps the theory of turbulence).

Part 4

Problems for Children 5 to 15 Years Old

Problems

I wrote these problems in Paris in the spring of 2004. Some Russian residents of Paris had asked me to help cultivate a culture of thought in their young children. This tradition in Russia far surpasses similar traditions in the West.

I am deeply convinced that this culture is developed best through early and independent reflection on simple, but not easy, questions, such as are given below. (I particularly recommend Problems 1, 3, and 13.)

My long experience has shown that C-level students, lagging in school, can solve these problems better than outstanding students, because the survival in their intellectual "Kamchatka" at the back of the classroom "demanded more abilities than are requisite to govern Empires", as Figaro said of himself in the Beaumarchais play. A-level students, on the other hand, cannot figure out "what to multiply by what" in these problems. I have even noticed that five year olds can solve problems like this better than can school-age children, who have been ruined by coaching, but who, in turn, find them easier than college students who are busy cramming at their universities. (And Nobel prize or Fields Medal winners are the worst at all in solving such problems.)

1. Masha was seven kopecks short of the price of an alphabet book, and Misha was one kopeck short. They combined their money to buy one book to share, but even then they did not have enough. How much did the book cost?

2. A bottle with a cork costs \$1.10, while the bottle alone costs 10 cents more than the cork. How much does the cork cost?

3. A brick weighs one pound plus half a brick. How many pounds does the brick weigh?

4. A spoonful of wine from a barrel of wine is put into a glass of tea (which is not full). After that, an equal spoonful of the (non-homogeneous) mixture from the glass is put back into the barrel. Now there is a certain volume of "foreign" liquid in each vessel (wine in the glass and tea in the barrel). Is the volume of foreign liquid greater in the glass or in the barrel?

5. Two elderly women left at dawn, one traveling from A to B and the other from B to A. They were heading towards one another (along the same road). They met at noon, but did not stop, and each of them kept walking at the same speed as before. The first woman arrived at B at 4 PM, and the second arrived at A at 9 PM. At what time was dawn on that day?

6. The hypotenuse of a right-angled triangle (on an American standardized test) is 10 inches, and the altitude dropped to it is 6 inches. Find the area of the triangle.

American high school students had been successfully solving this problem for over a decade. But then some Russian students arrived from Moscow, and none of them was able to solve it as their American peers had (by giving 30 square inches as the answer). Why not?

7. Victor has 2 more sisters than he has brothers. How many more daughters than sons do Victor's parents have?

8. There is a round lake in South America. Every year, on June 1, a Victoria Regia flower appears at its center. (Its stem rises from the bottom, and its petals lie on the water like those of a water lily). Every day the area of the flower doubles, and on July 1, it finally covers the entire lake, drops its petals, and its seeds sink to the bottom. On what date is the area of the flower half that of the lake?

9. A peasant must take a wolf, a goat and a cabbage across a river in his boat. However the boat is so small that he is able to take only one of the three on board with him. How can he transport all three across the river? (The wolf cannot be left alone with the goat, and the goat cannot be left alone with the cabbage.)

10. During the daytime a snail climbs 3cm up a post. During the night it falls asleep and slips down 2cm. The post is 10m high, and a delicious sweet is waiting for the snail on its top. In how many days will the snail get the sweet?

11. A hunter walked from his tent 10 km. south, then turned east, walked straight eastward 10 more km, shot a bear, turned north and after another 10 km found himself by his tent. What color was the bear and where did all this happen?

12. High tide occurred today at 12 noon. What time will it occur (at the same place) tomorrow?

13. Two volumes of Pushkin, the first and the second, are side-by-side on a bookshelf. The pages of each volume are 2cm thick, and the front and back covers are each 2mm thick. A bookworm has gnawed through (perpendicular to the pages) from the first page of volume 1 to the last page of volume 2. How long is the bookworm's track? [This topological problem with an incredible answer–4 mm–is totally impossible for academicians, but some preschoolers handle it with ease.]

14. Viewed from above and from the front, a certain object (a polyhedron) gives the shapes shown. Draw its shape as viewed from the side. (Hidden edges of the polyhedron are to be shown as dotted lines.)

15. How many ways are there to break the number 64 up into the sum of ten natural numbers, none of which is greater than 12? Sums which differ only in the order of the addends are not counted as different.

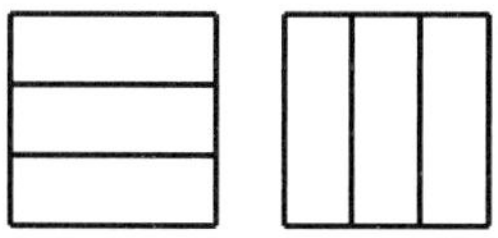

Top view Front view

To Problem 14

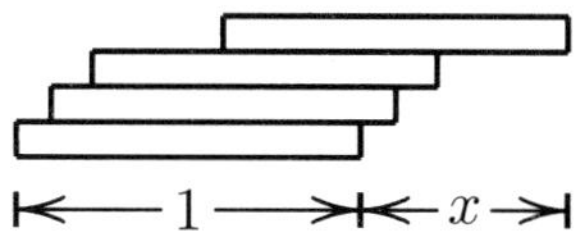

To Problem 16

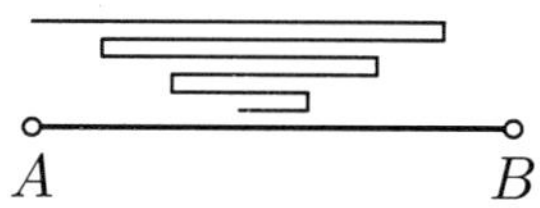

To Problem 17

16. We have a number of identical bars (say, dominoes). We want to stack them so that the highest hangs out over the lowest by a length equal to x bar-lengths. What is the largest possible value of x?

17. The distance between towns A and B is 40km. Two cyclists leave from A and B simultaneously traveling towards one another, one at a speed of 10km/h and the other at a speed of 15km/h. A fly leaves A together with the first cyclist, and flies towards the second at a speed of 100km/h. The fly reaches the second cyclist, touches his forehead, then flies back to the first, touches his forehead, returns to the second, and so on until the cyclists collide with their foreheads and squash the fly. How many kilometers has the fly flown altogether?

18. Vanya solved a problem about two pre-school age children. He had to find their ages (which are integers), given the product of their ages.

Vanya said that this problem could not be solved. The teacher praised him for a correct answer, but added to the problem the condition that the name of the older child was Petya. Then Vanya could solve the problem right away. Now you solve it.

19. Is the number 140359156002848 divisible by 4206377084?

20. One domino covers two squares of a chessboard. Cover all the squares except for its two opposite corners (on the same diagonal) with 31 dominoes. (A chessboard consists of $8 \times 8 = 64$ squares.)

21. A caterpillar wants to slither from the front left corner of the floor of a cubical room to the opposite corner (the right rear corner of the ceiling). Find the shortest route for such a journey along the walls of the room.

22. You have two vessels of volumes 5 liters and 3 liters. Measure out one liter, leaving the liquid in one of the vessels.

23. There are five heads and fourteen legs in a family. How many people and how many dogs are in the family?

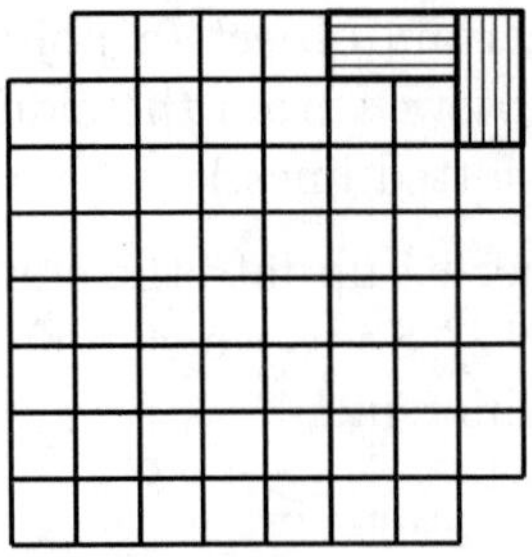

To Problem 20

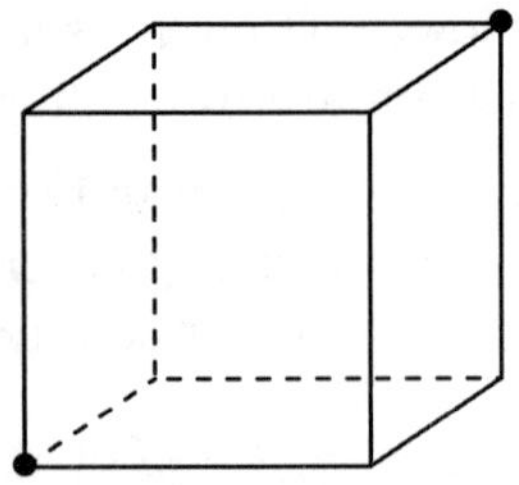

To Problem 21

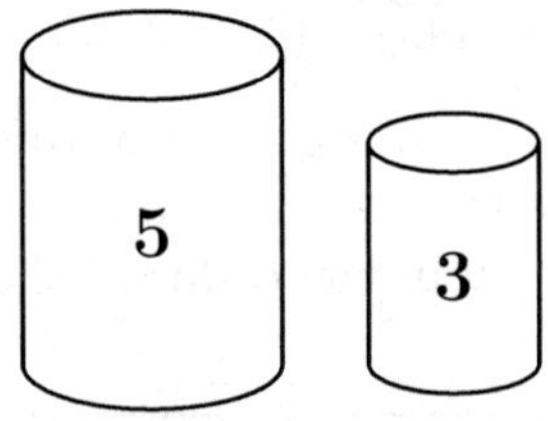

To Problem 22

24. Equilateral triangles are constructed externally on sides AB, BC, and CA of a triangle ABC. Prove that their centers (marked by asterisks on the diagram) form an equilateral triangle.

25. What polygons may be obtained as sections of a cube cut off by a plane? Can we get a pentagon? A heptagon? A regular hexagon?

26. Draw a straight line through the center of a cube so that the sum of the squares of the distances to it from the eight vertices of the cube is (a) maximal, (b) minimal (as compared with other such lines).

27. A right circular cone is cut by a plane along a closed curve. Two spheres inscribed in the cone are tangent to the plane, one at point A and the other at point B. Find a point C on the cross-section such that the sum of the distances $CA + CB$ is (a) maximal, (b) minimal.

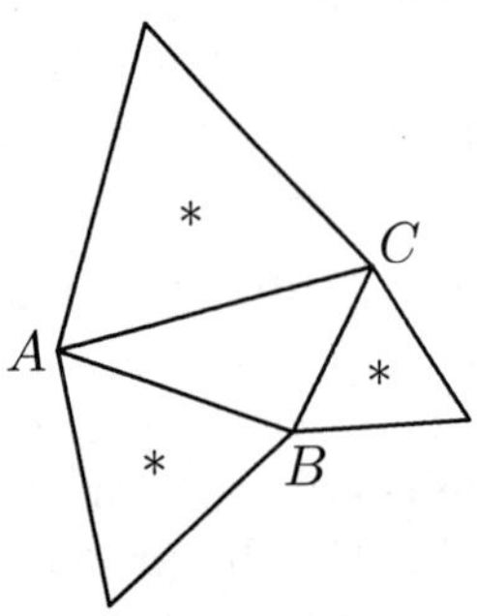

To Problem 24

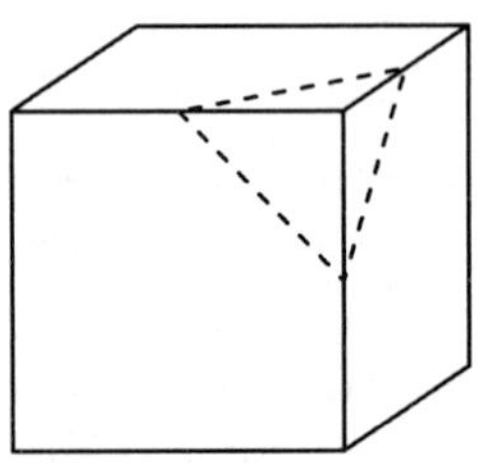

To Problem 25

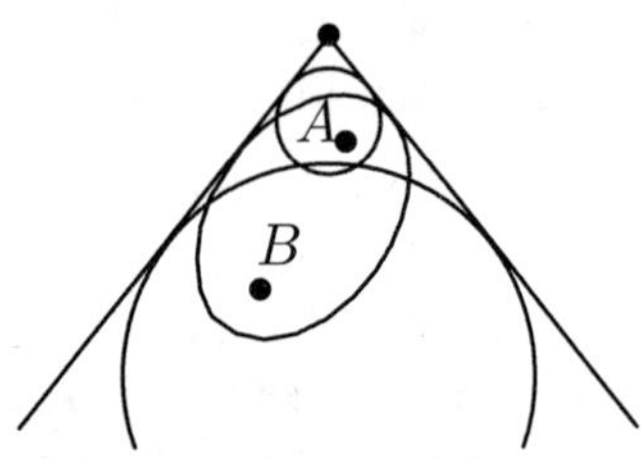

To Problem 27

28. The Earth's surface is projected onto a cylinder formed by the lines tangent to the meridians at the points where they intersect the equator. The projection is made along rays parallel to the plane of the equator and passing through the axis of the earth that connects its north and south poles. Will the area of the projection of France be greater or less than the area of France itself?

29. Prove that the remainder upon division of the number 2^{p-1} by an odd prime p is 1 (for example: $2^2 = 3a + 1$, $2^4 = 5b + 1$, $2^6 = 7c + 1$, $2^{10} - 1 = 1023 = 11 \cdot 93$).

30. A needle 10 cm. long is thrown randomly onto ruled paper. The distance between neighboring lines on the paper is also 10 cm. This is repeated N (say, a million) times. How many times (approximately, up to a few per cent error) will the needle fall so that it intersects a line on the paper?

One can perform this experiment with $N = 100$ instead of a million throws. (I did this when I was 10 years old.)

[The answer to this problem is surprising: $\frac{2}{\pi}N$. Moreover, even for a curved needle of length $a \cdot 10$cm. the number of intersections observed over N throws will be approximately $\frac{2a}{\pi}N$. The number $\pi \approx \frac{355}{113} \approx \frac{22}{7}$.]

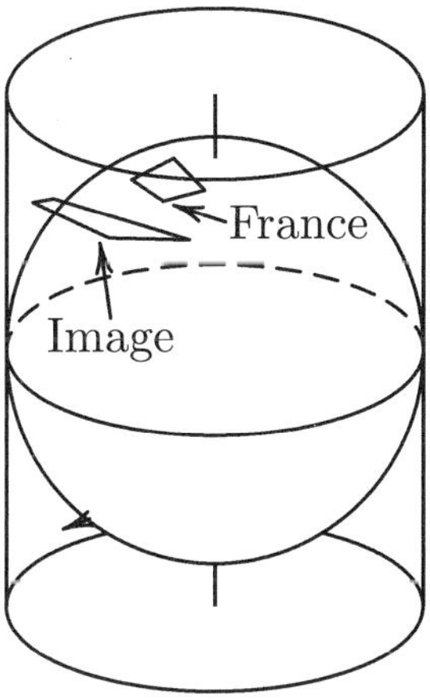

To Problem 28

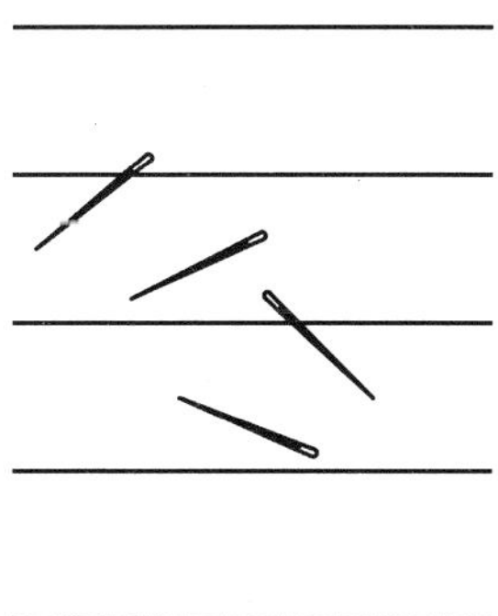

To Problem 30

31. Some polyhedra have only triangular faces. Some examples are the Platonic solids: the (regular) tetrahedron (4 faces), the octahedron (8 faces), and the icosahedron (20 faces). The faces of the icosahedron are all identical, it has 12 vertices, and it has 30 edges.

Is it true that for any such solid (a bounded convex polyhedron with triangular faces) the number of faces is equal to twice the number of vertices minus four?

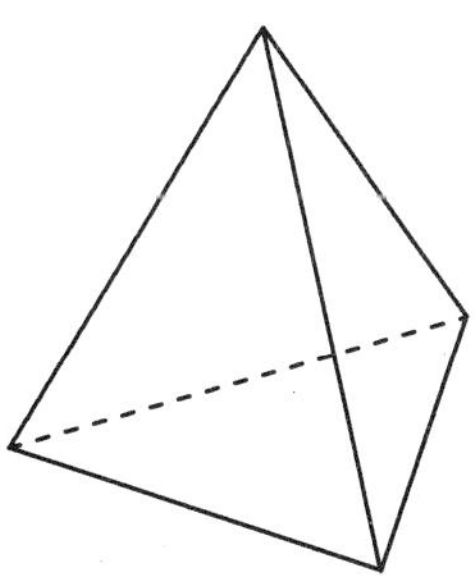

tetrahedron
(tetra = 4)

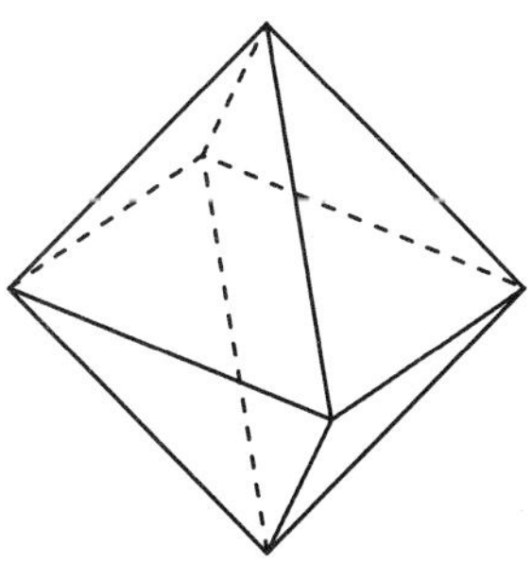

octahedron
(octa = 8)

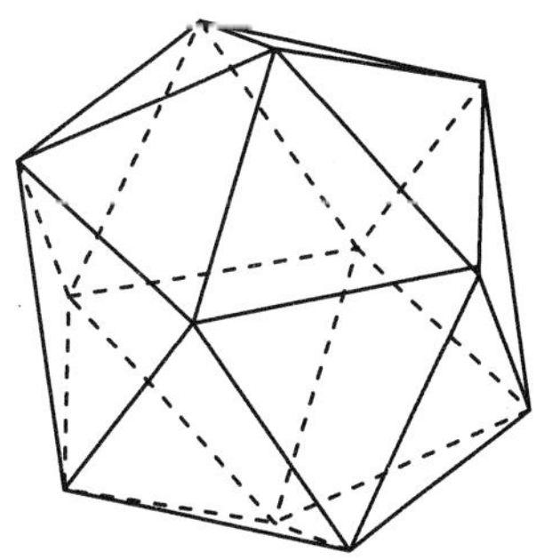

icosahedron
(icosa = 20)

32. There is one more Platonic solid (there are 5 of them altogether): a dodecahedron. It is a convex polyhedron with twelve (regular) pentagonal faces, twenty vertices and thirty edges (its vertices are the centers of the faces of an icosahedron).

Inscribe five cubes in a dodecahedron, whose vertices are also vertices of the dodecahedron, and whose edges are diagonals of faces of the dodecahedron. (A cube has 12 edges, one for each face of the dodecahedron). [This construction was invented by Kepler to describe his model of the planets.]

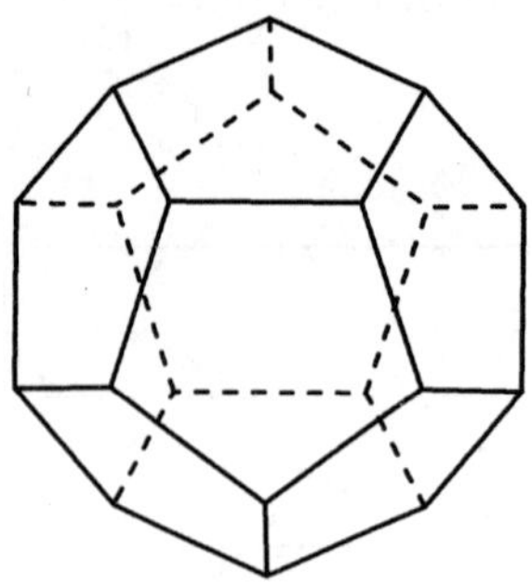

To Problem 32

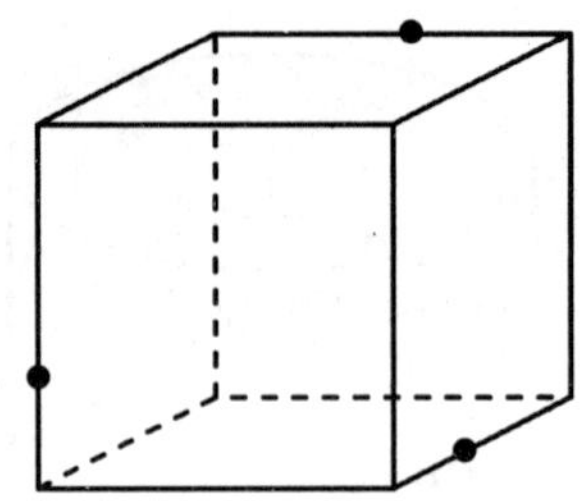

To Problem 33bis

33. Two regular tetrahedra can be inscribed in a cube, so that their vertices are also vertices of the cube, and their edges are diagonals of the cube's faces. Describe the intersection of these tetrahedra.

What fraction of the cube's volume is the volume of this intersection?

33bis. Construct the section cut of a cube cut off by the plane passing through three given points on its edges. [Draw the polygon along which the plane intersects the faces of the cube.]

34. How many symmetries does a tetrahedron have? A cube? An octahedron? An icosahedron? A dodecahedron? A symmetry of a figure is a transformation of this figure preserving lengths.

How many of these symmetries are rotations, and how many are reflections in planes (in each of the five cases listed)?

35. How many ways are there to paint the six faces of similar cubes with six colors (1,. . . ,6) [one color per face] so that no two of the colored cubes obtained are the same (that is, no two can be transformed into each other by a rotation)?

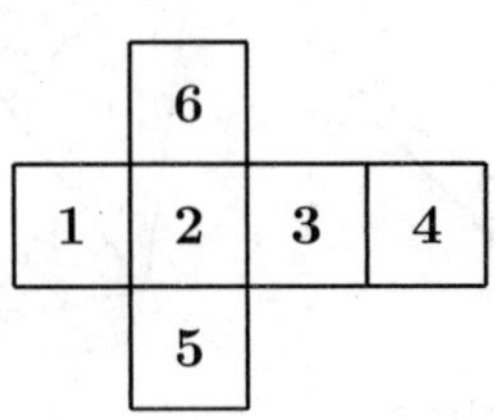

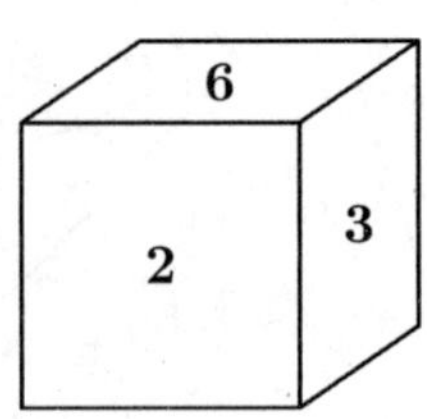

36. How many different ways are there to permute n objects?

For $n = 3$ there are six ways: $(1, 2, 3)$, $(1, 3, 2)$, $(2, 1, 3)$, $(2, 3, 1)$, $(3, 1, 2)$, $(3, 2, 1)$. What if the number of objects is $n = 4$? $n = 5$? $n = 6$? $n = 10$?

37. A cube has 4 major diagonals (that connect its opposite vertices). How many different permutations of these four objects are obtained by rotations of a cube?

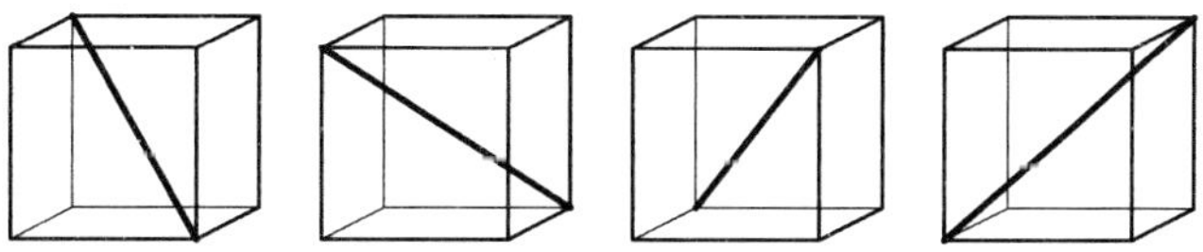

38. The sum of the cubes of several integers is subtracted from the cube of the sum of these numbers. Is this difference always divisible by 3?

39. Answer the same question for the fifth powers and divisibility by 5, and for the seventh powers and divisibility by 7.

40. Calculate the sum

$$\frac{1}{1 \cdot 2} + \frac{1}{2 \cdot 3} + \frac{1}{3 \cdot 4} + \cdots + \frac{1}{99 \cdot 100}$$

(with an error of not more than 1% of the correct answer).

41. If two polygons have equal areas, then they can be cut into a finite number of polygonal parts which may then be rearranged to obtain both the first and second polygons. Prove this. [For spatial solids this is not the case: the cube and the tetrahedron of equal volumes cannot be cut this way!]

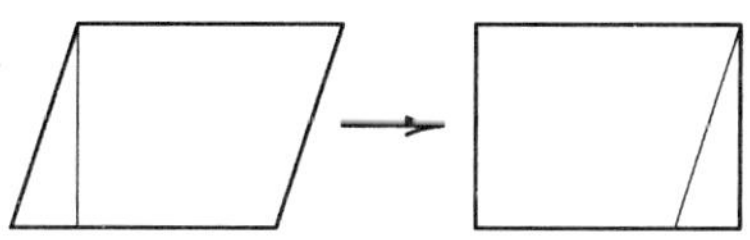

42. Four lattice points on a piece of graph paper are the vertices of a parallelogram. It turns out that there are no other lattice points either on the sides of the parallelogram or inside it. Prove that the area of such a parallelogram is equal to that of one of the squares of the graph paper.

43. Suppose, in Problem 42, there turn out to be a lattice points inside the parallelogram, and b lattice points on its sides. Find its area.

44. Is the statement analogous to the result of problem 43 true for parallelepipeds in 3-space?

45. The Fibonacci ("rabbit") numbers are the sequence ($a_1 = 1$), $1, 2, 3, 5, 8, 13, 21, 34, \ldots$, for which $a_{n+2} = a_{n+1} + a_n$ for any $n = 1, 2, \ldots$. Find the greatest common divisor of the numbers a_{100} and a_{99}.

46. Find the number of ways to cut a convex n-gon into triangles by cutting along non-intersecting diagonals. (These are the *Catalan* numbers, $c(n)$). For example, $c(4) = 2$, $c(5) = 5$, $c(6) = 14$. How can one find $c(10)$?

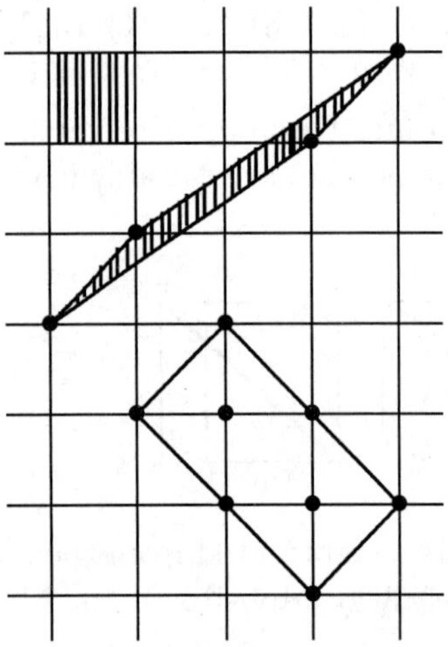

$a = 2, b = 2$

To Problems 42, 43

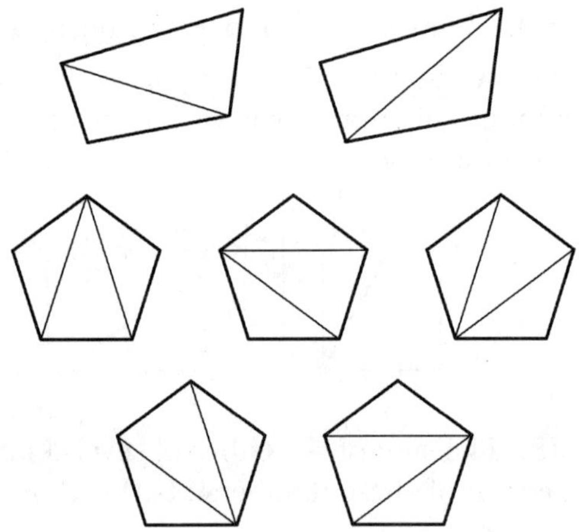

To Problem 46

47. There are n teams participating in a tournament. After each game, the losing team is knocked out of the tournament, and after $n-1$ games the team left is the winner of the tournament.

A schedule for the tournament may be written symbolically as (for example) $((a,(b,c)),d)$. This notation means that there are four teams participating. First b plays c, then the winner plays a, then the winner of this second game plays d.

How many possible schedules are there if there are 10 teams in the tournament?

For 2 teams, we have only (a,b), and there is only one schedule.

For 3 teams, the only possible schedules are $((a,b),c)$, or $((a,c),b)$, or $((b,c),a)$, and are 3 possible schedules.

For 4 teams we have 15 possible schedules:

$$(((a,b),c),d);\quad (((a,c),b),\ d);\quad (((a,d),b),c);\quad (((b,c),a),d);$$
$$(((b,d),a),c);\quad (((c,d),a),b);\quad (((a,b),d),c);\quad (((a,c),d),b);$$
$$(((a,d),c),b);\quad (((b,c),d),a);\quad (((b,d),c),a);\quad (((c,d),b),a);$$
$$((a,b),(c,d));\quad ((a,c),(b,d));\quad ((a,d),(b,c)).$$

48. We connect n points $1, 2, \ldots, n$ with $n-1$ segments to form a tree. How many different trees can we get? (Even the case $n = 5$ is interesting!)

$n = 2$: 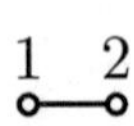

the number $= 1$;

$n = 3$: 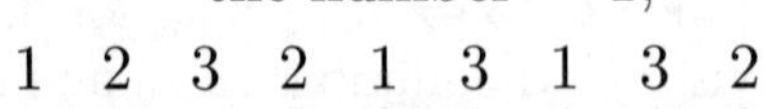

the number $= 3$;

49. A permutation $(x_1, x_2, \ldots, x_n)$ of the numbers $\{1, 2, \ldots, n\}$ is called a *snake* (of length n) if $x_1 < x_2 > x_3 < x_4 > \cdots$.

$n = 4$: 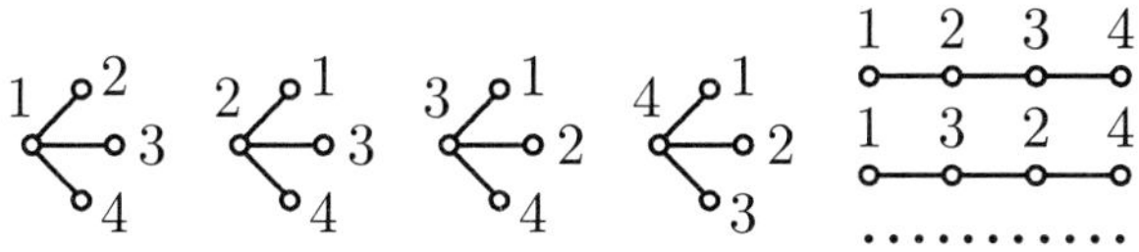 the number $= 16$.

Examples. $n = 2$, only $1 < 2$, the number $= 1$;

$n = 3$, $\left.\begin{array}{l} 1 < 3 > 2 \\ 2 < 3 > 1 \end{array}\right\}$, the number $= 2$;

$n = 4$, $\left.\begin{array}{l} 1 < 3 > 2 < 4 \\ 1 < 4 > 2 < 3 \\ 2 < 3 > 1 < 4 \\ 2 < 4 > 1 < 3 \\ 3 < 4 > 1 < 2 \end{array}\right\}$, the number $= 5$;

Find the number of snakes of length 10.

50. Let s_n denote the number of snakes of length n, so that

$$s_1 = 1,\ s_2 = 1,\ s_3 = 2,\ s_4 = 5,\ s_5 = 16,\ s_6 = 61.$$

Prove that the Taylor series for the tangent function is

$$\tan x = 1\frac{x^1}{1!} + 2\frac{x^3}{3!} + 16\frac{x^5}{5!} + \cdots = \sum_{k=1}^{\infty} s_{2k-1}\frac{x^{2k-1}}{(2k-1)!}.$$

51. Find the sum of the series

$$1 + 1\frac{x^2}{2!} + 5\frac{x^4}{4!} + 61\frac{x^6}{6!} + \cdots = \sum_{k=0}^{\infty} s_{2k}\frac{x^{2k}}{(2k)!}.$$

52. For $s > 1$, prove the identity

$$\prod_{p=2}^{\infty} \frac{1}{1 - \dfrac{1}{p^s}} = \sum_{n+1}^{\infty} \frac{1}{n^s}.$$

(The product is over all prime numbers p, and the summation over all natural numbers n.)

53. Find the sum of the series

$$1 + \frac{1}{4} + \frac{1}{9} + \cdots = \sum_{n=1}^{\infty} \frac{1}{n^2}.$$

$\left(\text{Prove that it is equal to } \dfrac{\pi^2}{6}, \text{ or approximately } \dfrac{3}{2}\right)$.

54. Find the probability that the fraction $\dfrac{p}{q}$ is in lowest terms.

This probability is defined as follows: in the disk $p^2 + q^2 \leq R^2$, we count the number $N(R)$ of points with integer coordinates p and q not having a common divisor greater than 1. Then we take the limit of the ratio $N(R)/M(R)$, where $M(R)$ is the total number of integer points in the disk ($M \sim \pi R^2$).

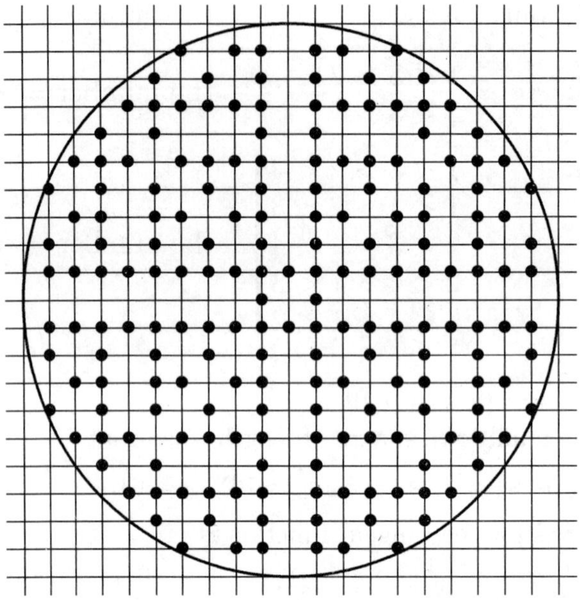

$N(10) = 192$
$M(10) = 316$
$N/M = 192/316$
≈ 0.6076

55. The sequence of Fibonacci numbers was defined in problem 45. Find the limit of the ratio a_{n+1}/a_n as n approaches infinity:

$$\frac{a_{n+1}}{a_n} = 2, \frac{3}{2}, \frac{5}{3}, \frac{8}{5}, \frac{13}{8}, \frac{21}{13}, \frac{34}{21}, \dots.$$

Answer: "The golden ratio", $\dfrac{\sqrt{5}+1}{2} \approx 1.618$.

This is the ratio of the sides of a postcard which stays similar to itself if we snip off a square whose side is the smaller side of the postcard.

How is the golden ratio related to a regular pentagon and a five-pointed star?

56. Calculate the value of the infinite continued fraction

$$1 + \cfrac{1}{2 + \cfrac{1}{1 + \cfrac{1}{2 + \cfrac{1}{1 + \cfrac{1}{\ddots}}}}} = a_0 + \cfrac{1}{a_1 + \cfrac{1}{a_2 + \cfrac{1}{a_3 + \cfrac{1}{\ddots}}}},$$

where $a_{2k} = 1$, $a_{2k+1} = 2$.

That is, find the limit as n approaches infinity of

$$a_0 + \cfrac{1}{a_1 + \cfrac{1}{a_2 + \cfrac{1}{\ddots + \cfrac{1}{a_n}}}}.$$

57. Find the polynomials $y = \cos 3(\arccos x)$, $y = \cos 4(\arccos x)$, $y = \cos n(\arccos x)$, where $|x| \leq 1$.

58. Calculate the sum of the kth powers of the n complex n^{th} roots of unity.

59. On the (x, y)−plane, draw the curves defined parametrically:

$$\{x = \cos 2t,\ y = \sin 3t\},\ \{x = t^3 - 3t,\ y = t^4 - 2t^2\}.$$

60. Calculate (with an error of not more than 10% of the answer)

$$\int_0^{2\pi} \sin^{100} x \, dx.$$

61. Calculate (with an error of not more than 10% of the answer)

$$\int_1^{10} x^x \, dx.$$

62. Find the area of a spherical triangle with angles (α, β, γ) on a sphere of radius 1. (The sides of such a triangle are great circles; that is, cross-sections of the sphere formed by planes passing through its center).

Answer: $S = \alpha + \beta + \gamma - \pi$. (For example, for a triangle with three right angles, $S = \pi/2$, that is, one-eighth of the total area of the sphere).

63. A circle of radius r rolls (without slipping) inside a circle of radius 1. Draw the whole trajectory of a point on the rolling circle (this trajectory is called a hypocycloid) for $r = 1/3$, $r = 1/4$ for $r = 1/n$, for $r = p/q$, and for $r = 1/2$.

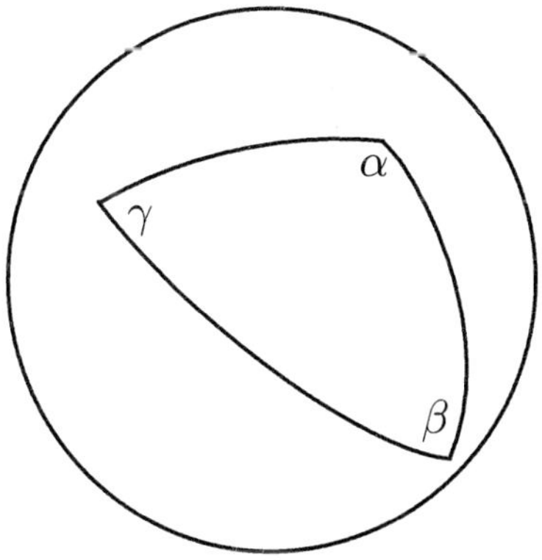

To Problem 62

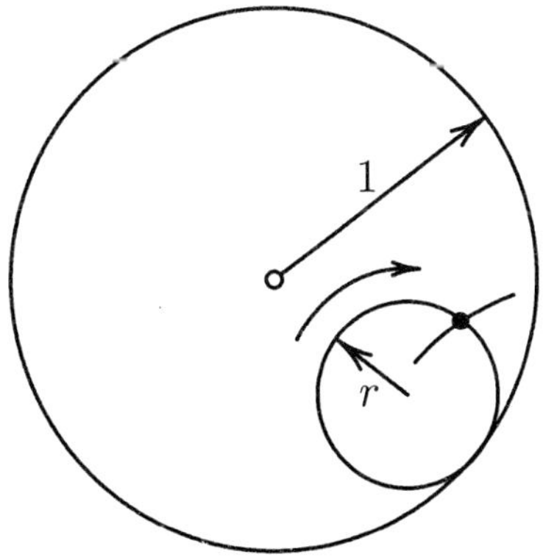

To Problem 63

64. In a class of n students, estimate the probability that two students have the same birthday. Is this a high probability? Or a low one?

Answer: (Very) high if the number of the pupils is (well) above some number n_0, (very) low if it is (well) below n_0, and what this n_0 actually is (when the probability $p \approx 1/2$) is what the problem is asking.

65. Snell's law states that the angle α made by a ray of light with the normal to layers of a stratified medium satisfies the equation $n(y) \sin \alpha =$ const, where $n(y)$ is the index of refraction of the layer at height y. (The quantity n is inversely proportional to the speed of light in the medium if we take its speed in a vacuum to be 1. In water $n = 4/3$).

Draw the rays forming the light's trajectories in the medium "air above a desert", where the index $n(y)$ has a maximum at a certain height. (See the diagram on the right.)

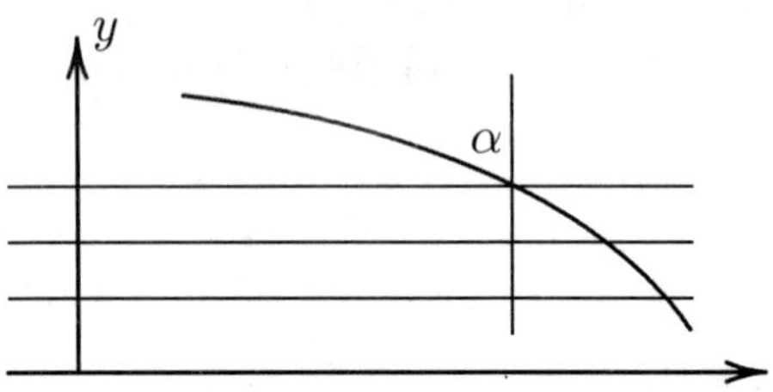

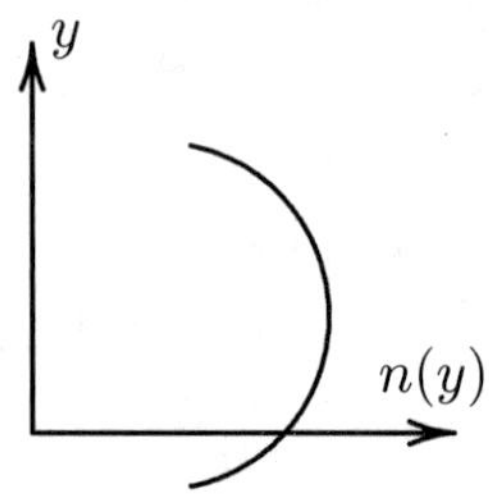

(A solution to this problem explains the phenomenon of mirages to those who understand how trajectories of rays emanating from objects are related to their images).

66. In an acute angled triangle ABC inscribe a triangle KLM of minimal perimeter (with its vertex K on AB, L on BC, M on CA).

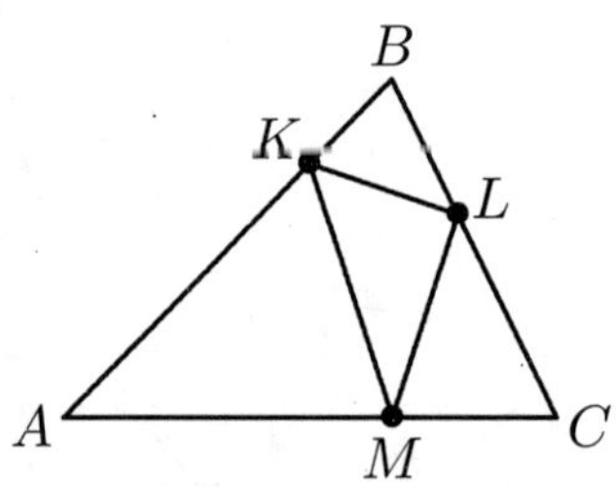

Hint: The answer for non-acute angled triangles is not nearly as beautiful as the answer for acute angled triangles.

67. Calculate the average value of the function $1/r$ (where $r^2 = x^2 + y^2 + z^2$ is the distance to the origin from the point with coordinates (x, y, z)) on the sphere of radius R centred at the point (X, Y, Z).

Hint: The problem is related to Newton's law of gravitation and Coulomb's law in electricity. In the two-dimensional version of the problem, the given function should be replaced by $\ln r$, and the sphere by a circle.

68. The fact that $2^{10} = 1024 \approx 10^3$ implies that $\log_{10} 2 \approx 0.3$. Estimate by how much they differ, and calculate $\log_{10} 2$ to three decimal places.

69. Find $\log_{10} 4$, $\log_{10} 8$, $\log_{10} 5$, $\log_{10} 50$, $\log_{10} 32$, $\log_{10} 128$, $\log_{10} 125$, and $\log_{10} 64$ with the same precision.

70. Using the fact that $7^2 \approx 50$, find an approximate value for $\log_{10} 7$.

71. Knowing the values of $\log_{10} 64$ and $\log_{10} 7$, find $\log_{10} 9, \log_{10} 3$, $\log_{10} 6, \log_{10} 27$, and $\log_{10} 12$.

72. Using the fact that $\ln(1+x) \approx x$ (where ln means $\log_e$), find $\log_{10} e$ and $\ln 10$ from the relation[16]

$$\log_{10} a = \frac{\ln a}{\ln 10}$$

and from the values of $\log_{10} a$ computed earlier (for example, for $a = 128/125$, $a = 1024/1000$ and so on).

Solutions to Problems 67–71 will give us, after a half hour of computation, a table of four-digit logarithms of any numbers using products of numbers whose logarithms have been already found as points of support and the formula

$$\ln(1+x) \approx x - \frac{x^2}{2} + \frac{x^3}{3} - \frac{x^4}{4} + \cdots$$

for corrections. (This is how Newton compiled a table of 40-digit logarithms!).

73. Consider the sequence of powers of two: 1, 2, 4, 8, 16, 32, 64, 128, 256, 512, 1024, 2048, ... Among the first twelve numbers, four have decimal numerals starting with 1, and none have decimal numerals starting with 7.

Prove that in the limit as $n \to \infty$ each digit will be met with as the first digit of the numbers 2^m, $0 \le m \le n$, with a certain average frequency: $p_1 \approx 30\%, p_2 \approx 18\%, \ldots, p_9 \approx 4\%$.

74. Verify the behavior of the first digits of powers of three: 1, 3, 9, 2, 8, 2, 7,.... Prove that, in the limit, here we also get certain frequencies and that the frequencies are same as for the powers of two. Find an exact formula for $p_1, \ldots,\ p_9$.

Hint: The first digit of a number x is determined by the fractional part of the number $\log_{10} x$. Therefore one has to consider the sequence of fractional parts of the numbers ma, where $a = \log_{10} 2$.

Prove that these fractional parts are uniformly distributed over the interval from 0 to 1: of the n fractional parts of the numbers ma, $0 \le m < n$,

---

[16]Euler's constant $e = 2.71828\cdots$ is defined as the limit of the sequence $\left(1+\frac{1}{n}\right)^n$ as $n \to \infty$. It is equal to the sum of the series $1+\frac{1}{1!}+\frac{1}{2!}+\frac{1}{3!}+\cdots$. It can also be defined by the given formula for $\ln(1+x)$: $\lim\limits_{x\to 0} \frac{\ln(1+x)}{x} = 1$.

a subinterval A will contain the quantity $k_n(A)$ such that as $n \to \infty$, $\lim(k_n(A)/n) =$ the length of the subinterval A.

75. Let $g : M \to M$ be a smooth map of a bounded domain M onto itself which is one-to-one and preserves areas (volumes in the multi-dimensional case) of domains.

Prove that in any neighborhood U of any point of M and for any N there exists a point x such that $g^T x$ is also in U for a certain integer $T > N$ (the "Recurrence Theorem").

76. Let M be the surface of a torus (with coordinates $\alpha \pmod{2\pi}, \beta \pmod{2\pi}$), and let $g(\alpha,\ \beta) = (\alpha + 1, \beta + \sqrt{2})$. Prove that for every point x of M the sequence of points $\{g^T(x)\}, T = 1, 2, \ldots$ is everywhere dense on the torus.

77. In the notation of problem 76, let

$$g(\alpha,\ \beta) = (2\alpha + \beta, \alpha + \beta) \pmod{2\pi}.$$

Prove that there is an everywhere dense subset of the torus consisting of periodic points x (that is, such that $g^{T(x)} = x$ for some integer $T(x) > 0$).

78. In the notation of Problem 77 prove that, for almost all points x of the torus, the sequence of points $\{g^T(x)\},\ T = 1, 2, \ldots$ is everywhere dense on the torus (that is, the points x without this property form a set of measure zero).

79. In Problems 76 and 78, prove that the sequence $\{g^T(x)\},\ T = 1, 2, \ldots$ is distributed over the torus uniformly: if a domain A contains $k_n(A)$ points out of the n points with $T = 1, 2, \ldots, n$, then

$$\lim_{n \to \infty} \frac{k_n(A)}{n} = \frac{\operatorname{mes} A}{\operatorname{mes} M}$$

(for example, for a Jordan measurable domain A of measure mes A).

Note to Problem 13. In posing this problem, I have tried to illustrate the difference in approaches to research by mathematicians and physicists in my invited paper in the journal "Advances in Physical Sciences" for the 2000 Centennial issue. My success far surpassed the goal I had in mind: the editors, unlike the preschool students on the experience with whom I based my plans, could not solve the problem. So they changed it to fit my answer of 4mm. in the following way: instead of "from the first page of the first volume to the last page of the second", they wrote "from the *last* page of the first volume to the *first* page of the second".

This true story is so implausible that I am including it here: the proof is the editors' version published by the journal.

Solutions to Selected Problems*

6. Such a triangle cannot exist: if the hypotenuse is 10 inches long, then the triangle can be inscribed into a half-disc of diameter 10 inches, and its altitude cannot exceed 5 inches.

8. On June 30.

10. Answer: after 998 days. Indeed, at the end of the first day the snail will be 3 centimeters high, at the end of the second day it will be 4 centimeters high, and so on. At the end of 998-th day the snail will be 10 meters high.

13. The answer 4 millimeters may seem unexpected, but look at Figure 41.

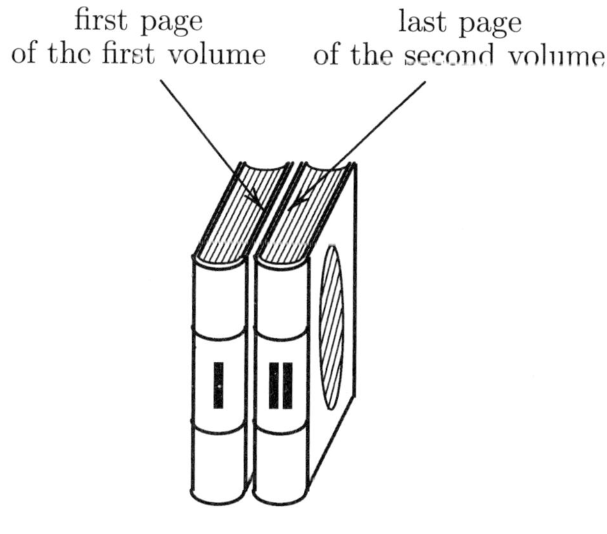

FIGURE 41. To Solution 13.

14. There are many such bodies. See an illustration (Figure 42) of one obtained from a cube by removing two triangular prisms. A side view is also shown.

15. Answer: 4,447. There are many different ways to count this number without listing all the partitions (although a computer program can do this in a fraction of a second). For example, one can use the following trick. Let

*Composed by Dmitry Fuchs

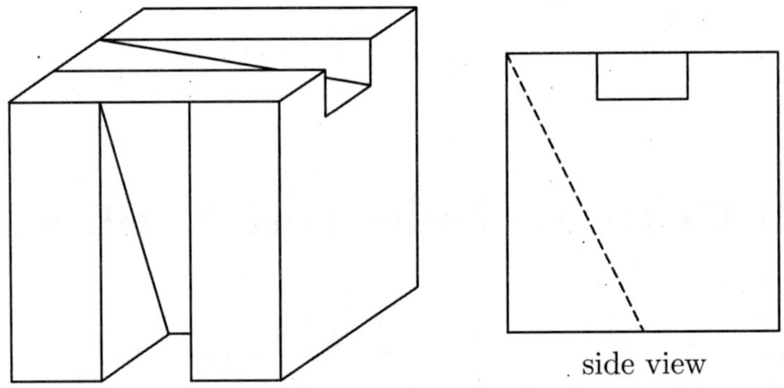

FIGURE 42. To Solution 14.

$P(n; m, k)$ be the number of partitions $n = a_1 + \cdots + a_k$ where the integers a_i satisfy the inequalities $m \geq a_1 \geq \cdots \geq a_k \geq 0$. What we need to find is $P(43; 11, 9)$. There is an obvious equality

$$P(n; m, k) = P(n; m, k-1) + P(n-k; m-1, k)$$

(which is obtained when we count separately partitions $m = a_1 + \cdots + a_k$ with $a_k = 0$ and $a_k \geq 1$). We apply it sufficiently many times:

$$\begin{aligned} P(43; 11, 9) &= P(43; 11, 8) + P(34; 10, 9) \\ &= P(43; 11, 7) + P(35; 10, 8) + P(34; 10, 8) + P(25; 9, 9) = \cdots \end{aligned}$$

(using, when necessary, the fact that $P(n; m, k) = 0$, if $n > mk$), and eventually we obtain

$$\cdots = 4447\, P(0; 0, 1) = 4447.$$

16. The length is unlimited. Indeed, consider a tower of $n+1$ identical plates of length 1, and introduce notations $x_1, \ldots, x_n$ as shown in Figure 43.

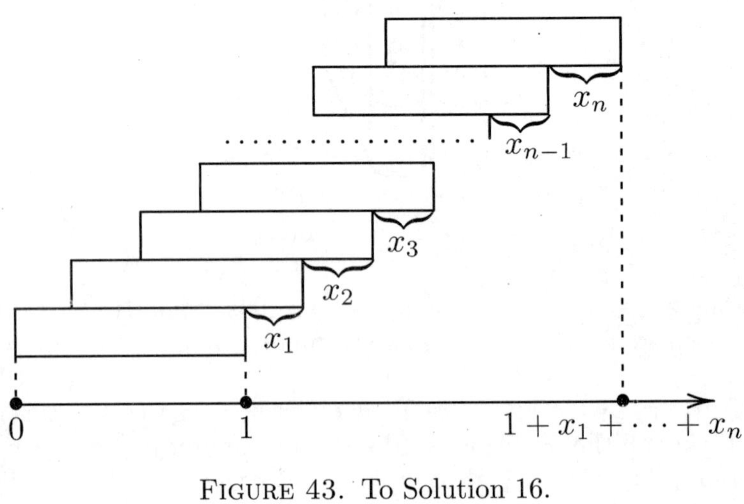

FIGURE 43. To Solution 16.

The length of the canopy is $x_1 + \cdots + x_n$.

For this tower being stable, it is needed that for $k = 1, \ldots, n$ the mass center of the union of plates ## $k+1, \ldots, n+1$ be located over some point of the plate #k. The horizontal coordinate of the mass center of the plate #k is $\frac{1}{2} + x_1 + \cdots + x_{k-1}$. Since all the plates are identical, the horizontal coordinate of the mass center of the union described is the mean of the horizontal coordinates of mass centers of the plates in this union, that is,

$$\frac{\left(\frac{1}{2} + x_1 + \cdots + x_k\right) + \cdots + \left(\frac{1}{2} + x_1 + \cdots + x_n\right)}{n-k+1}$$
$$= \frac{1}{2} + x_1 + \cdots + x_k + \frac{n-k}{n-k+1} x_{k+1} + \cdots + \frac{1}{n-k+1} x_n.$$

This last sum should be less than the horizontal coordinate of the right end of the plate #k, that is, less than $1 + x_1 + \cdots + x_{k-1}$. We arrive at the inequality

$$x_k + \frac{n-k}{n-k+1} x_{k+1} + \cdots + \frac{1}{n-k+1} x_n < \frac{1}{2},$$

that is,

$$(n-k+1)x_k + (n-k)x_{k+1} + \cdots + x_n < \frac{n-k+1}{2}. \qquad (*)$$

Let us assume that for all j, $x_j < \dfrac{1}{2(n-j+1)}$. Then each of the $n-k+1$ summands in the left hand side of the inequality $(*)$ is less that one half, and the inequality holds.

We see that the length of the canopy can be any number less than

$$\frac{1}{2n} + \frac{1}{2(n-1)} + \cdots + \frac{1}{4} + \frac{1}{2} = \frac{1}{2}\left(1 + \frac{1}{2} + \cdots + \frac{1}{n}\right).$$

It is well known that the sum in parentheses is unbounded, when n grows. Thus, the canopy may be arbitrarily long.

17. The cyclists met $\dfrac{40}{10+15} = 1.6$ hours after they started. Hence, the fly traveled $1.6 \cdot 100 = 160$ kilometers.

18. For Vanya, it was important to know not the name of the older boy, but rather the fact that the ages of the two boys were different.

19. The smaller number is divisible by 7, since 42, 63, 77, and 84 are divisible by 7. The bigger number is not divisible by 7, since 14, 35, 91, 56, and 28 are divisible by 7, but 48 is not divisible by 7. Therefore, the smaller number does not divide the bigger number.

FIGURE 44. To Solution 20.

20. Consider the standard coloring of the chess board (see Figure 44). It is obvious that a domino covers one white square and one black square. But the domain we want to cover contains unequal numbers of white and black squares (30 and 32). Therefore, it cannot be covered with non-overlapping dominoes.

21. The shortest path has length $a\sqrt{5}$ where a is the length of the edge of the cube. Moreover, there are 6 different paths of this length (see the left side of Figure 45).

First, any path can be shortened if it contains a connected non-straight part within any face of the cube: just replace this part by a straight interval with the same ends. This means that a shortest path must consist of straight intervals within the faces with both ends on edges. The starting point belongs to three faces, and the movement has to begin by a straight interval within one of them. This interval ends at one of the two edges opposite to the starting point, and this endpoint belongs to a face containing the endpoint. For shortness sake, we must go from this point straight to the endpoint. Thus, our shortest path belongs to the union of two adjacent faces. The right side of Figure 45 shows that, to be the shortest, the path must pass through the midpoint of the connecting edge.

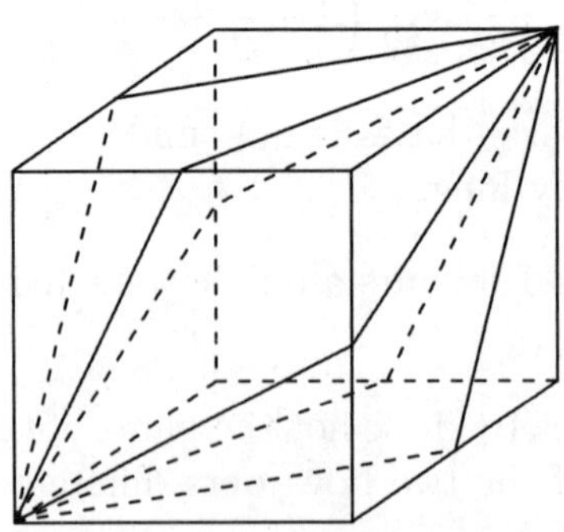

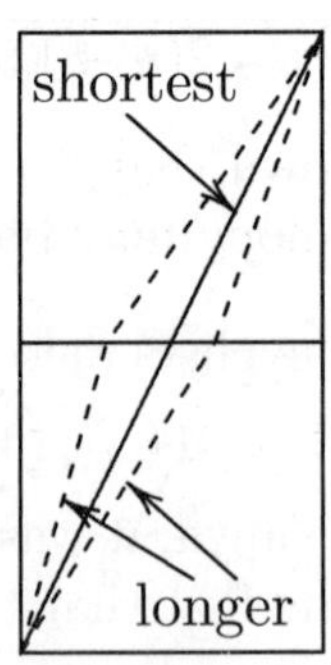

FIGURE 45. To Solution 21.

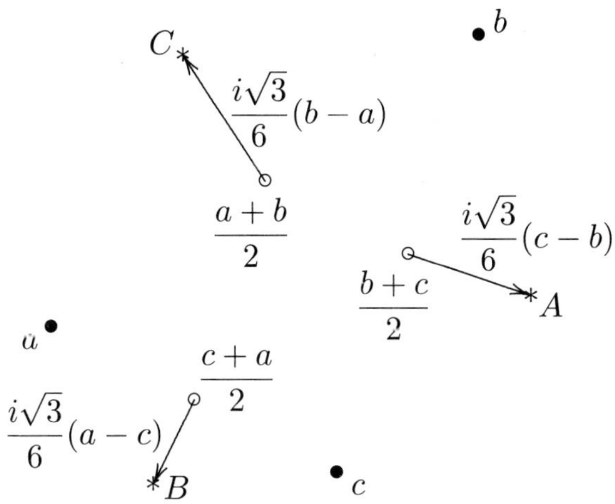

FIGURE 46. To Solution 24.

24. Let us think of the given triangle as drawn on the complex plane. Let a, b, c be complex numbers corresponding to its vertices. Then the midpoints of the sides will be $\frac{a+b}{2}, \frac{b+c}{2}, \frac{c+a}{2}$, and the vectors from these midpoints to the vertices of the new triangle, which we denote by A, B, C, will be obtained from the sides $b-a, c-b, a-c$ by counterclockwise rotation by $90°$ (that is, multiplication by i) and multiplication by $\frac{\sqrt{3}}{6}$ (which is one third of the ratio of lengths of an altitude and a side in an equilateral triangle). All this is shown in Figure 46 above.

Thus,

$$
\begin{aligned}
A &= \frac{b+c}{2} + \frac{i\sqrt{3}}{6}(c-b) = \frac{3-i\sqrt{3}}{6}b + \frac{3+i\sqrt{3}}{6}c,\\
B &= \frac{c+a}{2} + \frac{i\sqrt{3}}{6}(a-c) = \frac{3-i\sqrt{3}}{6}c + \frac{3+i\sqrt{3}}{6}a,\\
C &= \frac{a+b}{2} + \frac{i\sqrt{3}}{6}(b-a) = \frac{3-i\sqrt{3}}{6}a + \frac{3+i\sqrt{3}}{6}b,
\end{aligned}
$$

and

$$
\begin{aligned}
A-B &= -\frac{3+i\sqrt{3}}{6}a + \frac{3-i\sqrt{3}}{6}b + \frac{i\sqrt{3}}{3}c,\\
B-C &= -\frac{3+i\sqrt{3}}{6}b + \frac{3-i\sqrt{3}}{6}c + \frac{i\sqrt{3}}{3}a,\\
C-A &= -\frac{3+i\sqrt{3}}{6}c + \frac{3-i\sqrt{3}}{6}a + \frac{i\sqrt{3}}{3}b.
\end{aligned}
$$

To prove that ABC is an equilateral triangle, we need to check that $B - C$ is obtained from $A - B$ by a counterclockwise rotation by 120°, that is, by multiplication by $\cos 120° + i \sin 120° = \dfrac{-1 + i\sqrt{3}}{2}$; this is checked by an immediate calculation. (In the same way, we can check that $C - A = \dfrac{-1 + i\sqrt{3}}{2}(B - C)$ and $A - B = \dfrac{-1 + i\sqrt{3}}{2}(C - A)$, but we do not need this.)

25. The section is a polygon whose sides are intersections of the plane with faces. Since the cube has 6 faces, the number of sides of the polygon cannot exceed 6. For any $n \leq 6$ there is an n-gonal section (see Figure 47 below).

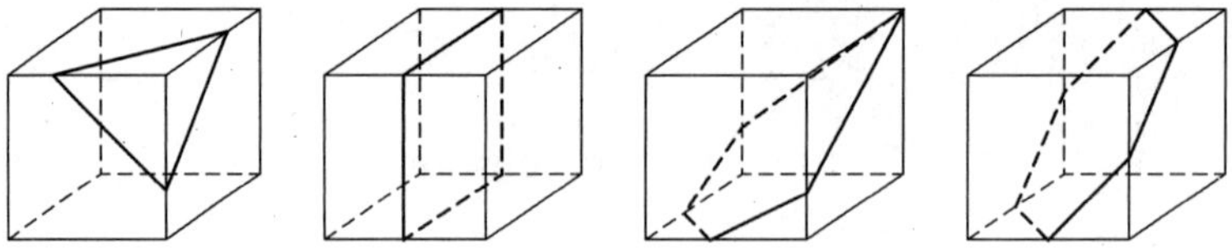

FIGURE 47. To Solution 25.

Hexagonal, quadrilateral, and triangular sections may be regular (shown in the picture); in particular, a regular hexagonal section can be obtained if (and only if) the plane passes through the center of the cube and is orthogonal to a big diagonal. A pentagonal section may appear when the plane passes through one of the vertices; it is never regular. With a certain abuse of language, we can say that 2-gonal section (an interval), 1-gonal section (a point), and 0-gonal section (the empty set) are also possible.

26. This sum does not depend on the line: it is always equal to $4\ell^2$ where ℓ is the length of the edge of the cube.

To prove this, let us first find the square of the distance from an arbitrary point (a, b, c) in space to an arbitrary line passing through the origin. The parametric equations of this line have the form $x = \alpha t, y = \beta t, z = \gamma t$ where we can assume that $\alpha^2 + \beta^2 + \gamma^2 = 1$.

Let $(\alpha t, \beta t, \gamma t)$ be the base of the perpendicular dropped from the point (a, b, c) to our line. Figure 48 shows that $t = \|(a, b, c)\| \cos \varphi$; thus, t is the dot product of the vectors (a, b, c) and (α, β, γ), that is, $t = a\alpha + b\beta + c\gamma$. Hence, the square of the distance from our point to our line is

$$\begin{aligned}&(a - \alpha t)^2 + (b - \beta t)^2 + (c - \gamma t)^2\\ &\qquad = a^2 + b^2 + c^2 - 2(a\alpha + b\beta + c\gamma)t + (\alpha^2 + \beta^2 + \gamma^2)t^2\\ &\qquad = a^2 + b^2 + c^2 - (a\alpha + b\beta + c\gamma)^2.\end{aligned}$$

Assume now that the eight vertices of the cube have coordinates $(\pm1, \pm1, \pm1)$. Then the center of the cube is the origin, and we can apply the previous formula. Our sum of the squares is

$$\sum\nolimits_{\pm}(3 - (\pm\alpha \pm \beta \pm \gamma)^2) = 24 - \sum\nolimits_{\pm}(\pm\alpha \pm \beta \pm \gamma)^2.$$

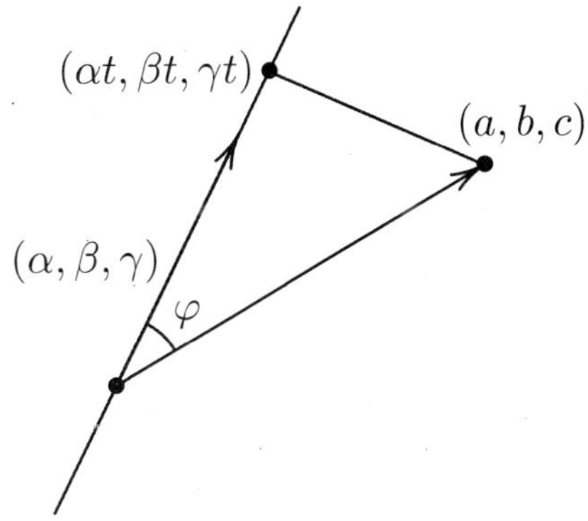

FIGURE 48. To Solution 26.

The last sum contains 8 times $\alpha^2 + \beta^2 + \gamma^2 = 1$, and each of the double product $2\alpha\beta, 2\alpha\gamma, 2\beta\gamma$ appears 8 times, 4 times with the sign + and four times with the sign −. Thus, the final result is

$$24 - 8 = 16 = 4\ell^2 \text{ (since } \ell = 2).$$

Remark. Similar statements hold for all regular polyhedra.

27. The sum $CA + CB$ does not depend on the choice of the point C. The following proof was, probably, known to Appolonius (third century BC).

In addition to the drawing accompanying the statement of this problem, let us show the circles of tangency of the spheres to the cone. Choose an arbitrary point C on the section, and draw a line through this point and the vertex of the cone. Let A', B' be the intersection (tangency) points of

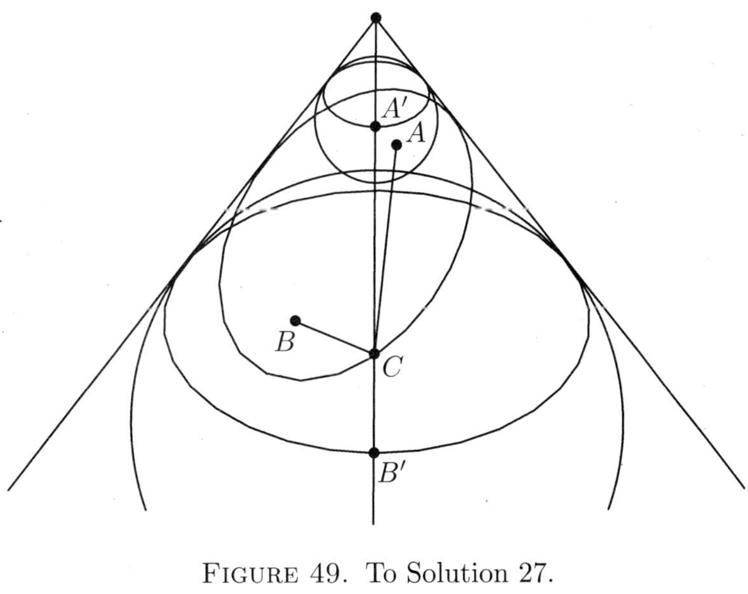

FIGURE 49. To Solution 27.

this line with the spheres. Then $CA = CA'$ (these are two tangents to a sphere from the same point), and similarly $CB = CB'$. Hence $CA + CB = CA' + CB' = A'B'$, and the latter, obviously, does not depend on the choice of C.

28. The area will be precisely equal to the area of France, and a similar thing holds for any area on the sphere. The following proof belongs to Cavalieri (1598–1647).

Take a point of the sphere, and let φ be the latitude of this point measured in radians $\left(\text{so } -\frac{\pi}{2} < \varphi < \frac{\pi}{2}\right)$. Then our projection onto the cylinder $\frac{1}{\cos\varphi}$ times stretches the parallel through this point and, in a proximity of this point, compresses the meridian (approximately) $\frac{1}{\cos\varphi}$ times (a picture below explains this).

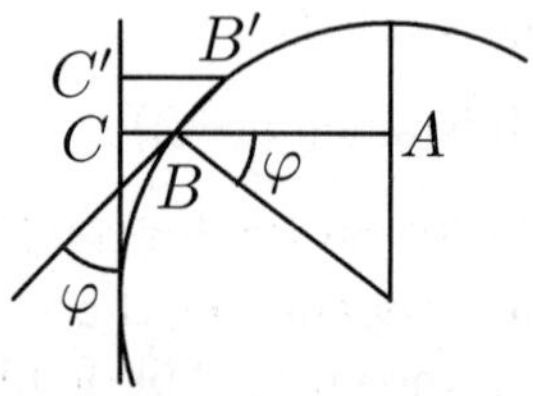

$$CC' = BB' \cos\varphi$$
$$AB = AC \cos\varphi$$

FIGURE 50. To Solution 28.

29. Use the binomial formula:

$$2^p = (1+1)^p = 1 + p + \frac{p(p-1)}{2} + \cdots + \frac{p(p-1)}{2} + p + 1.$$

The part of the last sum between the two 1's is divisible by p and is even (since 2^p is even). So, $2^p - 2 = Np$ where N is even. Divide the last equality by 2: $2^{p-1} - 1 = \frac{N}{2}p$. We see that $2^{p-1} - 1$ is divisible by p.

Remark. A more general result states that if p is prime and q is not divisible by p, then q^{p-1} has a remainder 1 upon division by p. This is called the *Little Fermat Theorem*.

30. It is not important that the length of the needle and the distance between the line are both 10 *cm*; all we need to know is that they are equal to each other. To make further formulas simpler, we assume that this length and this distance are both 2. Moving the needle horizontally does not change whether it intersects the line; so, we can assume that the midpoint of the needle is on a chosen vertical line. Similarly, moving the needle vertically by the distance of an even integer does not change whether it intersects the line; so, we can assume that the distance on the midpoint of the needle is between -1 and 1. With these assumptions, the position of the needle is described by two numbers: $h, -1 \le h \le 1$, which is the vertical coordinate

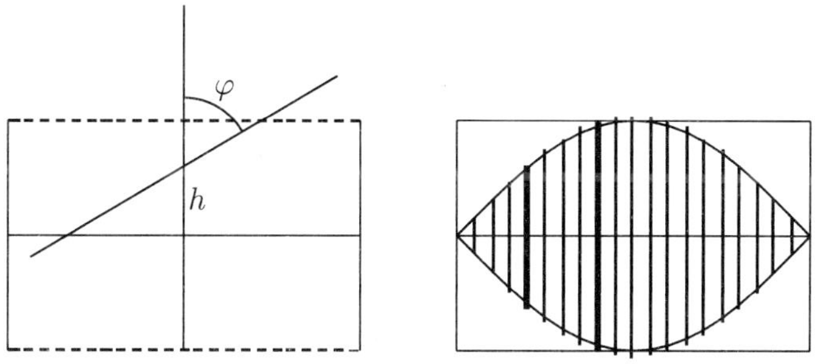

FIGURE 51. To Solution 30.

of the midpoint of the needle, and φ, $-\frac{\pi}{2} \le \varphi \le \frac{\pi}{2}$, the angle formed by the needle with the vertical direction (see the left side of Figure 51). We may assume that these two numbers are randomly chosen. It is clear that the needle may intersect only the line on this diagram (if the needle is vertical and $h = \pm 1$, then the needle also hits the other line, but this does not affect the probability). It is clear also that the needle intersects the line if and only if $\cos\varphi > |h|$.

In the plane (φ, h), the domain of all possible positions of the needle in the rectangle $-\frac{\pi}{2} \le \varphi \le \frac{\pi}{2}, -1 \le h \le 1$ of the area 2π and the domain of positions with an intersection with a line (shadowed in the right image of Figure 51) is bounded by the graphs $h = \pm\cos\varphi$ of the area 4 (it is proved by elementary calculus: $2\int_{-\pi/2}^{\pi/2} \cos\varphi\, d\varphi = 2\sin\varphi|_{-\pi/2}^{\pi/2} = 2(1-(-1)) = 4$). Thus, our probability is $\frac{4}{2\pi} = \frac{2}{\pi}$.

31. This follows from the famous Euler Theorem (proved, in fact, by Descartes 100 years before Euler) which states that if V, E, and F are numbers of vertices, edges, and faces of a convex polyhedron, then $V - E + F = 2$. If all the faces of the polyhedron are triangles, then $2E = 3F$. Indeed, let P be the number of all pairs (a face, an edge of this face); then $P = 3F$ and $P = 2E$. Thus, the Euler Theorem implies $V - \frac{3}{2}F + F = 2$ which becomes, after the multiplication by 2, $2V - 3F + 2F = 4$, that is, $2V = F + 4$.

32. In each of the twelve faces of the dodecahedron, we choose one the (five) diagonals. The first one we choose in an arbitrary way (in an arbitrarily chosen face), and then choose the rest of them using the following rule: if two faces are adjacent to each other, then the chosen diagonals either are both parallel to the common edge, or make different angles with the common face. See Figure 52.

The diagonals chosen form a cube, since any rotation of the dodecahedron which takes one of the chosen diagonals into another one takes the whole family of the chosen diagonals into itself.

There are five such cubes, because every face has five diagonals.

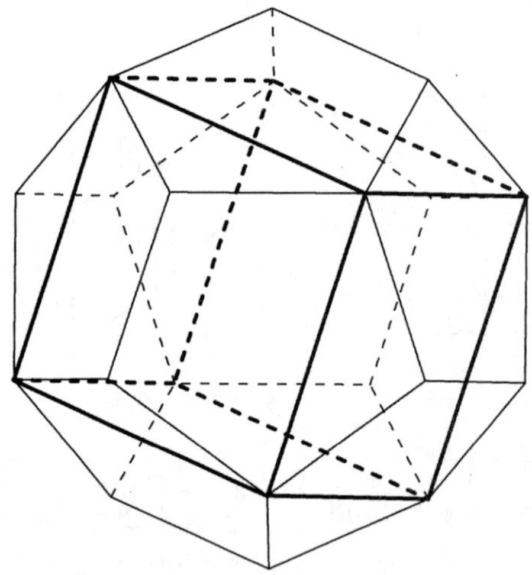

FIGURE 52. To Solution 32.

33. The intersection of the two tetrahedra is an octahedron whose six vertices are the centers of the faces of the cube (see Figure 53). Indeed, the two diagonals of the every face of the cube are edges of two different tetrahedra. So, their crossing point, which is the center of the cube, belongs to the intersection of the two tetrahedra. Consequently, the convex hull of the six-point set of the centers of the faces of the cube, which is the octahedron described, is contained in the intersection of the tetrahedra. To prove that this intersection contains nothing else, we notice that the eight triangular faces of the octahedron are contained in the eight triangular faces of the tetrahedra as triangles formed by the midpoints of the edges of the faces of the tetrahedra.

To evaluate the volume of the octahedron, we notice that the octahedron is the union of two quadrilateral pyramids, with the (square) base whose area is one half of the area of the face of the cube, and whose altitude is one half of the edge of the cube. Thus, the volume of each pyramid is $\frac{1}{3}\cdot\frac{1}{2}\cdot\frac{1}{2}=\frac{1}{12}$ of the volume of the cube, and the volume of the whole octahedron is $\frac{1}{6}$ of the volume of the cube.

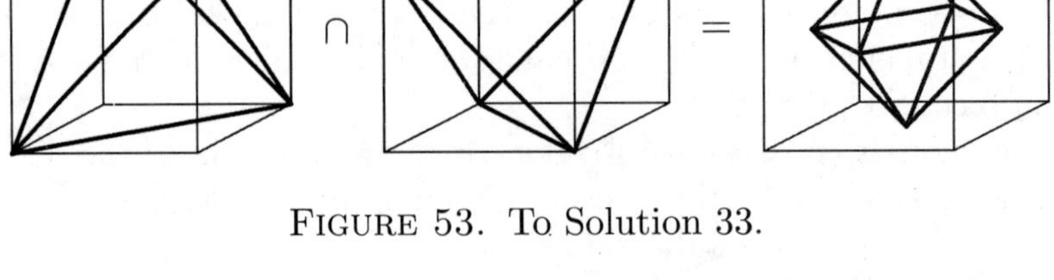

FIGURE 53. To Solution 33.

34. Let our regular polyhedron have a vertices, and let the number of edges converging at each vertex be b. Then the total number of symmetries is $2ab$, and ab of them are rotations. Indeed, fix a vertex A of our polyhedron. Then A can be taken by a rotational symmetry of the polyhedron into any other vertex B, and if B is specified, then all the symmetries which take A into B can be obtained from one of them by b rotations and b reflections. Thus:

– for a tetrahedron, there are $2 \cdot 4 \cdot 3 = 24$ symmetries, 12 of which are rotations;

– for a cube, there are $2 \cdot 8 \cdot 3 = 48$ symmetries, 24 of which are rotations;

– for an octahedron, there are $2 \cdot 6 \cdot 4 = 48$ symmetries, 24 of which are rotations;

– for an icosahedron, there are $2 \cdot 12 \cdot 5 = 120$ symmetries, 60 of which are rotations;

– for a dodecahedron, there are $2 \cdot 20 \cdot 3 = 120$ symmetries, 60 of which are rotations.

The number of reflections in planes is equal to the number of planes of symmetry. It is 6 for a tetrahedron, 9 for a cube and an octahedron, and 15 for an icosahedron and a dodecahedron.

35. There are 30 ways. Indeed, if we do not allow rotations, then there are $6 \cdot 5 \cdot \ldots 1 = 720$ colorings (we enumerate the faces of the cube by numbers $1, 2, \ldots, 6$ in an arbitrary way, then choose one of 6 colors for the face #1, one of 5 remaining colors for the face #2, and so on). No rotation (which is not the identity) takes any coloring into itself. There are 24 rotations of the cube (see Problem 34). So, the whole set of 720 colorings falls into the union of sets of rotationally equivalent colorings, each contains 24 colorings. Thus, up to a rotation, there are $720/24 = 30$ different colorings.

36. There are $n! = 1 \cdot 2 \cdot 3 \cdot \ldots \cdot n$ ways. In particular, for $n = 4, 5, 6, 10$, there are $24, 120, 720, 3628800$ ways.

37. Every symmetry of the cube yields a permutation of the four diagonals, and every permutation of the diagonals corresponds to two different symmetries. [Indeed, there are two symmetries, which take every diagonal into itself: the identity and the antipodal map (the reflection in the center). Hence, every permutation of the diagonals corresponds to two symmetries, which are obtained from each other by a composition with the antipodal map.] Precisely one of these two symmetries is a rotation. Thus, there is a one-to-one correspondence between rotations of the cube and permutation of the diagonals.

38. It is true, and it can be proven by induction with respect to the number n of integers. If $n = 1$, then the difference is 0, it is divisible by 3. Assume that for the sum of $n - 1$ integers the statement is true. Let $a_1, \ldots, a_n$ be the given integers, and let $b = a_1 + \cdots + a_{n-1}$. Then

$$(a_1 + \cdots + a_n)^3 = (b + a_n)^3 = b^3 + 3b^2 a_n + 3b a_n^2 + a_n^3,$$

and we see that

$$(a_1 + \cdots + a_n)^3 - b^3 - a_n^3$$

is divisible by 3. Hence

$$(a_1+\cdots+a_n)^3-a_1^3-\cdots-a_n^3 = [(a_1+\cdots+a_n)^3-b^3-a_n^3]+[b^3-a_1^3-\cdots-a_{n-1}^3],$$

and of two summands in square brackets, the first is divisible by 3 by the statement above, and the second is divisible by 3 by the induction hypothesis.

39. It is true, and the proof is almost the same as in the previous solution. The only difference is that we need to replace the formula for $(b + a_n)^3$ by one of the formulas

$$(b + a_n)^5 = b^5 + 5b^4a_n + 10b^3a_n^2 + 10b^2a_n^3 + 5ba_n^4 + a_n^5,$$
$$(b + a_n)^7 = b^7 + 7b^6a_n + 21b^5a_n^2 + 35b^4a_n^3 + 35b^3a_n^4 + 21b^2a_n^5 + 7ba_n^6 + a_n^7.$$

Similar arguments show that we can replace 3 (or 5, or 7) by any prime number, but for a composite exponent this may be not true: $(1+1)^4 - 1^4 - 1^4 = 14$ is not divisible by 4, $(1 + 1)^6 - 1^6 - 1^6 = 62$ is not divisible by 6.

40. It is

$$\left(1 - \frac{1}{2}\right) + \left(\frac{1}{2} - \frac{1}{3}\right) + \cdots + \left(\frac{1}{99} - \frac{1}{100}\right) = 1 - \frac{1}{100} = \frac{99}{100}$$

precisely (no error).

41. It is sufficient to prove that every polygon P can be cut by straight lines into pieces, of which one can assemble a rectangle of size $1 \times \text{area}(P)$. This can be done in five steps. (See Figure 54.)

Step One. Any polygon can be cut into several triangles. This allows us to assume that the polygon P itself is a triangle.

Step Two. We cut the triangle into two pieces by a line through midpoints of two sides, and from these pieces we assemble a parallelogram.

Step Three. We cut a triangle off the parallelogram by a line AB where A is a vertex, B is a point on a side opposite to A such that the length r of AB is rational. Then we attach this triangle to the other side of the parallelogram. As a result, we obtain a different parallelogram with one of the sides being of a rational length.

Step Four. Cut the parallelogram by parallel lines perpendicular to the edge AB of rational length r at the distance r from each other; two of these lines pass through A and B. Then rearrange the pieces to assemble a rectangle with one of the edges being AB.

Step Five. Let $r = \dfrac{p}{q}$. Divide the side AB into q equal pieces and divide the perpendicular side into p equal pieces. Then divide the rectangle into pq equal small rectangles by lines parallel to the sides. These pq small rectangles can be rearranged into a new rectangle with one side of length 1.

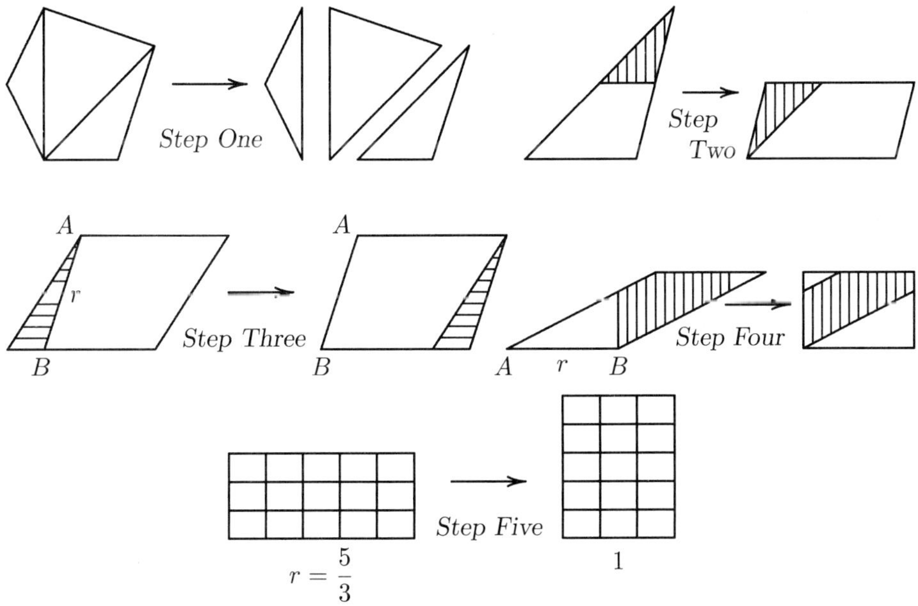

FIGURE 54. To Solution 41.

42. See Solution to Problem 43.

43. The area is $a + \frac{b}{2} + 1$. In the situation of Problem 42, $a = b = 0$, so the area equals 1, as stated.

Let s be the area of our parallelogram and $c = a + \frac{b}{2} + 1$; we want to prove that $s = c$.

Consider the tiling of the plane by parallelograms obtained from our parallelogram by shifting by the vectors $m_1 s_1 + m_2 s_2$ where s_1 and s_2 are the sides (considered as vectors) and m_1, m_2 are integers (see Figure 55).

Choose one of the four vertices of our parallelogram and call it the *prime vertex*. Then every tile of our tiling has a prime vertex. These prime vertices form a sublattice of the standard lattice (formed by the vertices of the graph paper). Fix some point in the plane and denote by $d(R)$ the (closed) disc centered at this point. Let $N(R)$ be the number of points of the standard lattice contained in $d(R)$, and let $M(R)$ be the number of points of our sublattice contained in $d(R)$.

Let us denote by $S(R)$ the union of tiles whose prime vertices lie in $d(R)$, and denote by ℓ the length of the longer diagonal of our parallelogram (we assume that the length of the side of the cell of the graph paper is 1). Then area, $S(R) = sM(R)$. Obviously, $d(R - \ell) \subset S(R) \subset d(R + \ell)$, so

$$\pi(R - \ell)^2 \leq sM(R) \leq \pi(R + \ell)^2. \tag{1}$$

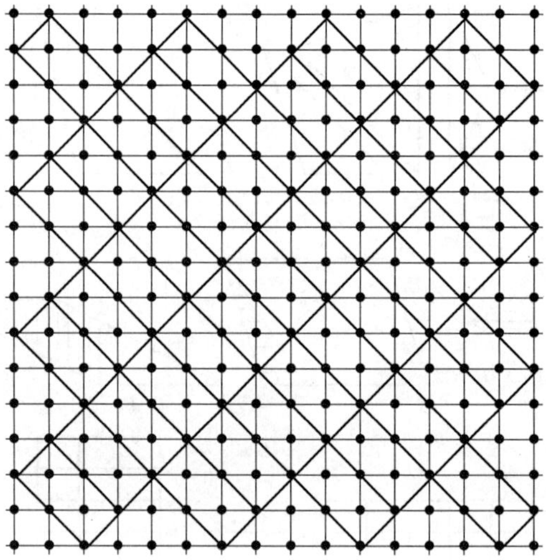

FIGURE 55. To Solution 43.

The same arguments applied to the tiling by the cells of the graph paper yield the inequalities

$$\pi(R-\sqrt{2})^2 \le N(R) \le \pi(R+\sqrt{2})^2 \tag{2}$$

(since the area of a cell is 1 and the length of the diagonal of a cell is $\sqrt{2}$).

Next, let us consider the product $cM(R)$. This can be regarded as the sum over all the tiles T in $S(R)$ where the summand corresponding to T is, in turn, the sum over all the vertices of the graph paper within T of summands equal to 1 for points inside the T, to $\frac{1}{2}$ for points inside the edges of T, and to $\frac{1}{4}$ for the vertices of T. This shows that the total contribution of a vertex of the graph paper to $cM(R)$ never exceeds 1, it is 1 for the vertices in $d(R-2\ell)$, and it can be positive only for vertices in $d(R+\ell)$. Thus

$$N(R-2\ell) \le cM(R) \le N(R+\ell), \tag{3}$$

which gives, in combination with (2),

$$\pi(R-2\ell-\sqrt{2})^2 \le cM(R) \le \pi(R+\ell+\sqrt{2})^2. \tag{4}$$

From (1) and (4), we can deduce for $\frac{c}{s} = \frac{cM(R)}{sM(R)}$

$$\frac{(R-2\ell-\sqrt{2})^2}{(R+\ell)^2} \le \frac{c}{s} \le \frac{(R+\ell+\sqrt{2})^2}{(R-\ell)^2}. \tag{5}$$

Since both the first and the third fractions in (5) become arbitrarily close to 1 when R grows, (5) shows that $\frac{c}{s}$ must be 1.

44. Yes, and the statement of Problem 43 as well. The latter means that if P is a parallelepiped in space whose vertices all have integer coordinates, and if a, b, and c are the number of points with integer coordinates, respectively, inside P, inside the faces of P, and inside the edges of P, then

$$\text{volume}\,(P) = a + \frac{b}{2} + \frac{c}{4} + 1.$$

If there are no points with integer coordinates in P (including the boundary) besides the vertices, that is, if $a = b = c = 0$, then the volume of P is 1. The proof is a replica of the proof in Solution to Problem 43. Similar facts hold in any dimension.

45. For any positive integers a, b, it is true that $(a+b, b) = (a, b)$. Using this, we have:

$$\begin{aligned}(a_{100}, a_{99}) &= (a_{98} + a_{99}, a_{99}) = (a_{98}, a_{99}) = (a_{99}, a_{98})\\ &= (a_{97} + a_{98}, a_{98}) = (a_{97}, a_{98}) = (a_{98}, a_{97})\\ &\ldots\ldots\ldots\ldots\ldots\ldots\ldots\ldots\ldots\ldots\ldots\ldots\\ &= (a_1 + a_2, a_2) = (a_1, a_2) = (1, 1) = 1.\end{aligned}$$

46. There is a recursion formula for $c(n)$ (in this formula, by definition, $c(2) = 1$; we can try to justify it by saying that a two-gon is divided by diagonals into the union of 0 triangles in one way). The formula:

$$c(n) = c(2)c(n-1) + c(3)c(n-2) + c(4)c(n-3) + \cdots + c(n-1)c(2).$$

To prove that, we first choose one of the sides of the n-gon. Then, for every partition of the n-gon into $n-2$ triangles there is a triangle containing the chosen side; to specify it, we need to choose one of the $n-2$ vertices not belonging to the chosen side. Figure 56 shows how it looks for a hexagon (the chosen side is the bottom side). If we remove the chosen side, then our n-gon falls into the union of an m-gon and an $(n+1-m)$-gon (with $m = 2, 3, \ldots, n-1$). To complete the partition of the n-gon into triangles by diagonals, we need to do this for both the m-gon and the $(n+1-m)$-gon, which can be done in $c(m)c(n+1-m)$ ways (for a fixed m). Whence our formula.

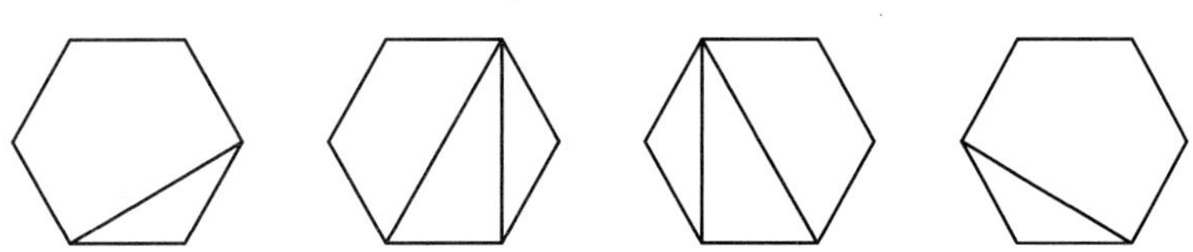

FIGURE 56. To Solution 46.

Now we use our formula for computations:

$$\begin{aligned}
&c(2) = 1,\\
&c(3) = 1\cdot 1 = 1,\\
&c(4) = 1\cdot 1 + 1\cdot 1 = 2,\\
&c(5) = 1\cdot 2 + 1\cdot 1 + 2\cdot 1 = 5,\\
&c(6) = 1\cdot 5 + 1\cdot 2 + 2\cdot 1 + 5\cdot 1 = 14,\\
&c(7) = 1\cdot 14 + 1\cdot 5 + 2\cdot 2 + 5\cdot 1 + 14\cdot 1 = 42,\\
&c(8) = 1\cdot 42 + 1\cdot 14 + 2\cdot 5 + 5\cdot 2 + 14\cdot 1 + 42\cdot 1 = 132,\\
&c(9) = 1\cdot 132 + 1\cdot 42 + 2\cdot 14 + 5\cdot 5 + 14\cdot 2 + 42\cdot 1 + 132\cdot 1 = 429,\\
&c(10) = 1\cdot 429 + 1\cdot 132 + 2\cdot 42 + 5\cdot 14 + 14\cdot 5\\
&\qquad\qquad + 42\cdot 2 + 132\cdot 1 + 429\cdot 1 = 1430.
\end{aligned}$$

Remark. The Catalan numbers have the property to appear in many very different combinatorial contexts. An interested reader can get familiar with some of them by Wikipedia. This source contains also an explicit (not recursion) formula for these numbers:

$$c(n) = \frac{(2n-4)!}{(n-1)!(n-2)!}.$$

47. The number of different schedules for a tournament of $n \geq 2$ teams is

$$1\cdot 3\cdot 5\cdot \ldots \cdot (2n-3)$$

(the product of all odd numbers from 1 to $2n-3$). Let us prove this by induction. If $n=2$, then there is only one schedule, which agrees with our formula. Assume that the result holds for $n-1$ teams, that is, there are $1\cdot 3\cdot 5\cdot \ldots \cdot (2n-5)$ schedules for $n-1$ teams. Imagine that the n-th team entered the tournament after the schedule for the $n-1$ teams had been already established. To include the new team, we need to do one of two things: either choose one of the $n-2$ games of the existing schedule and have the new team to play with one of the participants of the chosen game and then have the winner to play with the other participant (for this, we have $2(n-2)$ options); or have the new team play with the winner of the last game (one option for this). Thus a schedule for $n-1$ teams may be turned into a schedule for n teams in $2(n-2)+1 = 2n-3$ ways, and the total number of schedules for n teams is

$$[1\cdot 3\cdot 5\cdot \ldots \cdot (2n-5)]\cdot (2n-3),$$

which completes our induction.

48. The answer is n^{n-2}. This fact is called the *Cayley formula* and has several known proofs, neither of which is short and elementary. Below, we restrict ourselves to proving a recursion formula for the number of trees.

Let T_n be the number of trees with vertices $1, 2, \ldots, n$. Consider the set of such trees *with one edge marked*. The number of elements of such set is, obviously, $(n-1)T_n$. Let us count the number of elements of this set in a different way. First, we choose a marked edge; for this, there are

$\binom{n}{2} = \dfrac{n(n-1)}{2}$ options. If we remove, from a tree, the marked edge (but not its endpoints!), then our tree falls into a disjoint union of two trees, with the numbers of vertices m and $n-m$ (where $m = 0, 1, 2, \ldots, n$). For the trees, there are T_m and T_{n-m} options, but also we need to specify the vertices of, say, the first tree, for which there are $\binom{n-2}{m-1}$ options. Thus, the whole amount of options is $\sum_{m=0}^{n} \binom{n-2}{m-1} T_m T_{n-m}$, and we obtain the formula

$$(n-1)T_n = \frac{n(n-1)}{2} \sum_{m=0}^{n} \binom{n-2}{m-1} T_m T_{n-m},$$

or

$$2T_n = n \sum_{m=0}^{n} \binom{n-2}{m-1} T_m T_{n-m}. \qquad (1)$$

In particular,

$$\begin{aligned}
T_1 &= 1,\\
2T_2 &= 2(1\cdot 1) = 2, \quad T_2 = 1,\\
2T_3 &= 3(1\cdot 1 + 1\cdot 1) = 6, \quad T_3 = 3,\\
2T_4 &= 4(1\cdot 3 + 2\cdot 1\cdot 1 + 3\cdot 1) = 32, \quad T_4 = 16,\\
2T_5 &= 5(1\cdot 16 + 3\cdot 1\cdot 3 + 3\cdot 3\cdot 1 + 1\cdot 16\cdot 1) = 250, \quad T_5 = 125,\\
2T_6 &= 6(1\cdot 125 + 4\cdot 1\cdot 16 + 6\cdot 3\cdot 3 + 4\cdot 16\cdot 1 + 125\cdot 1) = 2592, \quad T_6 = R1296.
\end{aligned}$$

We leave to the reader to check that the numbers $T_n = n^{n-2}$ satisfy the relation (1).

49. For a snake of length n, mark the term x_k equal to n (see Figure 57).

Obviously, k must be even; indeed, if k is odd, then x_k must be less than at least one of x_{k-1} and x_{k+1} (whichever exists), which is impossible, if $x_k = n$. To the left and to the right of x_k, there are snakes of lengths $k-1$ and $n-k$, but (1) the left one should be read from the right to the left and (2) these snakes are permutations not of the set $\{1, 2, \ldots, n\}$, but of two complementary subsets of the set $\{1, 2, \ldots, n-1\}$ consisting, respectively, of $k-1$ and $n-k$ elements. Thus, for a snake of the length n, we need to specify an even number k, $1 \le k \le n$, and for this k, a set of $k-1$ numbers between 1 and $n-1$ (there are $\binom{n-1}{k-1}$ options for that), and two snakes of lengths $k-1$ and $n-k$ ($s(k-1)$ and $s(n-k)$ options, respectively). We arrive at the recursion formula

$$s(n) = \sum_{\substack{1\le k\le n\\ k \text{ is even}}} \binom{n-1}{k-1} s(k-1)s(n-k) \qquad (1)$$

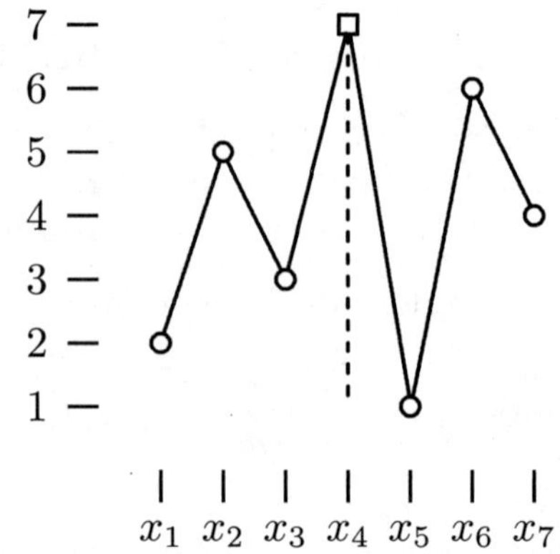

FIGURE 57. To Solution 49.

(where we assume $s(0) = s(1) = 1$). From this:

$$
\begin{aligned}
s(2) &= 1,\\
s(3) &= 2s(1)s(1) = 2,\\
s(4) &= 3s(1)s(2) + s(3)s(0) = 5,\\
s(5) &= 4s(1)s(3) + 4s(3)s(1) = 16,\\
s(6) &= 5s(1)s(4) + 10s(3)s(2) + s(5)s(0) = 61,\\
s(7) &= 6s(1)s(5) + 20s(3)s(3) + 6s(5)s(1) = 232,\\
s(8) &= 7s(1)s(6) + 35s(3)s(4) + 21s(5)s(2) + s(7)s(0) = 1345,\\
s(9) &= 8s(1)s(7) + 56s(3)s(5) + 56s(5)s(3) + 8s(7)s(1) = 7296,\\
s(10) &= 9s(1)s(8) + 72s(3)s(6) + 126s(5)s(4) + 36s(7)s(2) + s(9)s(0) = 46617.
\end{aligned}
$$

50. The function $y = \tan x$ satisfies the differential equation

$$y' = y^2 + 1,$$

and this, together with the condition $y(0) = 0$, uniquely determines the function $y = \tan x$. Since $y = \tan x$ is an odd function, its Taylor series involves only odd powers of x. Let

$$\tan x = \sum_{k=1}^{\infty} \frac{a_k}{(2k-1)!} x^{2k-1};$$

we want to prove that $a_k = s(2k-1)$. The derivation and squaring the power series above shows that

$$(\tan x)' = \sum_{k=1}^{\infty} \frac{a_k}{(2k-2)!} x^{2k-2},$$

$$(\tan x)^2 = \sum_{k=2}^{\infty} \left[\sum_{\substack{p+q=k \\ p\geq 1, q\geq 1}} \frac{a_p a_q}{(2p-1)!(2q-1)!} \right] x^{2k-2}.$$

From the differential equation, $a_1 = 1$ and, for $k \geq 2$,

$$\frac{a_k}{(2k-2)!} = \sum_{\substack{p+q=k \\ p\geq 1, q\geq 1}} \frac{a_p a_q}{(2p-1)!(2q-1)!}$$

or

$$a_k = \sum_{\substack{p+q=k \\ p\geq 1, q\geq 1}} \binom{2k-2}{2p-1} a_p a_q.$$

If we plug in this formula $a_k = s(2k-1)$, we will obtain precisely the recursion formula (1) from Solution to Problem 49 for $n = 2k-1$. Thus the numbers a_k satisfy the recursion formula for $s(2k-1)$, so $a_k = s(2k-1)$.

51. Let $f(x)$ the sum of the given series. Then

$$f'(x) = \sum_{k=1}^{\infty} \frac{s(2k)}{(2k-1)!} x^{2k-1},$$

and by formula (1) from Solution to Problem 49 this sum is equal to

$$\sum_{k=1} \sum_{p+q=k} \binom{2k-1}{2p-1} \frac{1}{(2k-1)!} s(2p-1) s(2q) x^{2k-1}$$

$$= \sum_{p=1}^{\infty} \sum_{q=0}^{\infty} \frac{s(2p-1)}{(2p-1)!} x^{2p-1} \cdot \frac{s(2q)}{(2q)!} x^{2q}.$$

Taking into account the result of Problem 50, we arrive at the differential equation

$$f'(x) = f(x) \cdot \tan x,$$

which, together with the condition $f(0) = 1$, uniquely determines $f(x)$. It is easy to check that the function $f(x) = \dfrac{1}{\cos x}$ satisfies the equation and the condition, so this is the sum of the series.

52. Let $p_1 < p_2 < p_3 < \cdots$ be the sequence of all primes. We have:

$$\prod_{k=1}^{\infty} \frac{1}{1-\dfrac{1}{p_k^s}} = \prod_{k=1}^{\infty} \sum_{m_k=0}^{\infty} \frac{1}{p_k^{sm_k}} = \sum_{m_1=0}^{\infty} \sum_{m_2=0}^{\infty} \sum_{m_3=0}^{\infty} \cdots \frac{1}{p_1^{sm_1} p_2^{sm_2} p_3^{sm_3} \cdots}$$
$$= \sum_{m_1=0}^{\infty} \sum_{m_2=0}^{\infty} \sum_{m_3=0}^{\infty} \cdots \frac{1}{(p_1^{m_1} p_2^{m_2} p_3^{m_3} \cdots)^s} = \sum_{n=1}^{\infty} \frac{1}{n^s}.$$

53. There are many different proofs of this fact, which was first observed (but not proved rigorously) by Euler. The reader can find them on Wikipedia. Euler's heuristic arguments were as follows. The function $f(x) = \dfrac{\sin x}{x}$ (equal, by definition, to 1 at 0) has zeroes at all points $x = n\pi, n \neq 0$ and has no other zeroes (even in the complex domain). We can expect that

$$\frac{\sin x}{x} = \cdots \left(1+\frac{x}{2\pi}\right)\left(1+\frac{x}{\pi}\right)\left(1-\frac{x}{\pi}\right)\left(1-\frac{x}{2\pi}\right)\cdots$$
$$= \left(1-\frac{x^2}{\pi^2}\right)\left(1-\frac{x^2}{4\pi^2}\right)\left(1-\frac{x^2}{9\pi^2}\right)\cdots$$

(since the right hand side has the same zeroes as $\dfrac{\sin x}{x}$ and equals 1 at 0; actually, the equality can be proved by instruments of complex analysis). The last product, turned into series, has the form

$$1 - \left(\sum_{n=1}^{\infty} \frac{1}{n^2\pi^2}\right) x^2 \pm \cdots;$$

compare this with the Taylor expansion

$$\frac{\sin x}{x} = 1 - \frac{1}{6}x^2 \pm \cdots$$

to get the equality

$$\sum_{n=1}^{\infty} \frac{1}{n^2\pi^2} = \frac{1}{6} \quad \text{or} \quad \sum_{n=1}^{\infty} \frac{1}{n^2} = \frac{\pi^2}{6}.$$

Remark. We can use the formulas obtained above to compute $\sum_{n-1}^{\infty} \dfrac{1}{n^s}$ for all even s. For example, let us do it for $s = 4$. Comparing the coefficients at x^4 in the above infinite product formula for $\dfrac{\sin x}{x}$ and using the Taylor expansion $\dfrac{\sin x}{x} = 1 - \dfrac{1}{6}x^2 + \dfrac{1}{120}x^4 \pm \cdots$, we get

$$\sum_{1 \le p < q < \infty} \frac{1}{p^2 q^2 \pi^4} = \frac{1}{120} \quad \text{or} \quad \sum_{1 \le p < q < \infty} \frac{1}{p^2 q^2} = \frac{\pi^4}{120}.$$

Furthermore, squaring the formula for $\sum_{n=1}^{\infty} \frac{1}{n^2} = \frac{\pi^2}{6}$, we get

$$\left(\sum_{n=1}^{\infty} \frac{1}{n^2}\right)^2 = \sum_{n=1}^{\infty} \frac{1}{n^4} + 2 \sum_{1 \le p < q < \infty} \frac{1}{p^2 q^2} = \frac{\pi^4}{36},$$

from which

$$\sum_{n=1}^{\infty} \frac{1}{n^4} = \frac{\pi^4}{36} - 2\frac{\pi^4}{120} = \frac{\pi^4}{90}.$$

Similar arguments yield formulas $\sum_{n=1}^{\infty} \frac{1}{n^6} = \frac{\pi^6}{945}$, $\sum_{n=1}^{\infty} \frac{1}{n^8} = \frac{\pi^8}{9450}$, and so on. Actually, $\sum_{n=1}^{\infty} \frac{1}{n^{2s}} = \rho_s \pi^{2s}$, where ρ_s is a rational number which can be found explicitly in terms of the so called *Bernoulli numbers*. However, no formula exists for $\sum_{n=1}^{\infty} \frac{1}{n^t}$ with t odd.

54. To do this, we will need the results of the two previous problems. The probability of the fact that an integer n is not divisible by some p is $1 - \frac{1}{p}$; the probability of the fact that two integers, m and n, do not share a divisor p, is $1 - \frac{1}{p^2}$. A fraction $\frac{m}{n}$ is not cancellable if the integers m and n do not share any prime divisors. Since these events for different primes are clearly independent, the probability of the fact that the fraction $\frac{m}{n}$ is not cancellable is $\prod_p \left(1 - \frac{1}{p^2}\right)$, where the product is taken over all primes. By the statement of Problem 52, the inverse to this product is $\sum_{n=1}^{\infty} \frac{1}{n^2}$, and by the statement of Problem 53, the last (infinite) sum equals $\frac{\pi^2}{6}$. Thus, our probability is $\frac{6}{\pi^2} \approx 0.608$.

55. The problem becomes easy if one assumes that the limit

$$r = \lim_{n \to \infty} \frac{a_{n+1}}{a_n}$$

exists. With this assumption, we take the equality

$$\frac{a_{n-1}}{a_n} + 1 = \frac{a_{n+1}}{a_n}$$

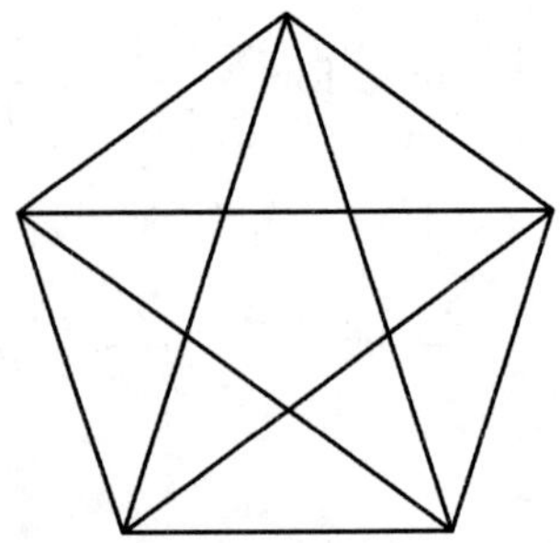

FIGURE 58. To Solution 55.

and apply to both sides $\lim_{n\to\infty}$. We get: $\frac{1}{r}+1=r$, that is, $r^2-r-1=0$, and this quadratic equation has only one positive root: the golden ratio $\tau=\frac{1+\sqrt{5}}{2}$.

We leave the proof of the existence of the limit to the reader. One of the ways: first prove (by induction) that $a_{n+1}a_{n-1}-a_n^2=(-1)^n$, and then notice that the sequence

$$d_n=\frac{a_{n+1}}{a_n}-\frac{a_n}{a_{n-1}}=\frac{a_{n+1}a_{n-1}-a_n^2}{a_na_{n-1}}=\frac{(-1)^n}{a_na_{n-1}}$$

has alternating signs and $\lim|d_n|=0$; this implies the existence of our limit.

As to the geometric question, the ratio of the lengths of the side and the diagonal of a regular pentagon (which is, simultaneously, the edge of the five-point star inscribed into the pentagon; see Figure 58) is equal to the golden ratio. To prove this, we need to know that $\cos 36^\circ=\frac{\tau}{2}$. This follows, in turn, from the equality $\cos(3\cdot 36^\circ)=-\cos(2\cdot 36^\circ)$. This can be regarded as a cubic equation for $\cos 36^\circ$, whose roots are $-1, \frac{1\pm\sqrt{5}}{4}$.

56. Again, we leave it to the reader to prove the existence of the limit. If it exists and is equal to r, then

$$r=1+\frac{1}{2+\frac{1}{r}}\Rightarrow r=\frac{3r+1}{2r+1}\Rightarrow 2r^2-2r-1=0.$$

The positive solution of this quadratic equation is $\frac{1+\sqrt{3}}{2}$ which is the answer to our problem.

Remark. The reader who finds the two last problems interesting may want to read Part 1, "Continued Fractions", of this volume, especially the section about the Lagrange Theorem.

57. To do this, we need to know the formulas for sines and cosines of multiple angles (which, certainly, are important and useful by themselves).

Namely, there are polynomials $P_n(u)$ and $Q_n(u)$ such that

$$\sin nx = \sin(x) \cdot Q_n(\cos x) \text{ and } \cos nx = P_n(\cos x).$$

For example, $P_1(u) = u, Q_1(u) = 1, P_2(u) = 2u^2 - 1, Q_2(u) = 2u$. If we know these polynomials, our problem is solved immediately: $\cos(n \arccos x) = P_n(x)$.

Using the relations

$$\begin{aligned} \sin(n+1)x &= \sin nx \cos x + \sin x \cos nx, \\ \cos(n+1)x &= \cos nx \cos x - \sin nx \sin x, \end{aligned}$$

we obtain recursion formulas for the polynomials P_n and Q_n,

$$\begin{aligned} P_{n+1}(u) &= uP_n(u) + (u^2 - 1)Q_n(u), \\ Q_{n+1}(u) &= uQ_n(u) + P_n(u), \end{aligned}$$

from which we can get the following table of polynomials P_n and Q_n.

n	$P_n(u)$	$Q_n(u)$
1	u	1
2	$2u^2 - 1$	$2u$
3	$4u^3 - 3u$	$4u^2 - 1$
4	$8u^4 - 8u^2 + 1$	$8u^3 - 4u$
5	$16u^5 - 20u^2 + 5u$	$16u^4 - 12u^2 + 1$
6	$32u^6 - 48u^4 + 18u^2 - 1$	$32u^5 - 32u^3 + 6u$
7	$64u^7 - 112u^5 + 56u^3 - 7u$	$64u^6 \quad 80u^4 + 24u^2 - 1$
...	...	...

This table suggests general formulas for $P_n(u)$ and $Q_n(u)$:

$$P_n(u) = \frac{1}{2} \sum_{j \le n/2} (-1)^j \frac{n}{n-j} \binom{n-j}{j} (2u)^{n-2j},$$

$$Q_n(u) = \sum_{j<(n-1)/2} (-1)^j \binom{n-1-j}{j-1} (2u)^{n-1-2j}.$$

It is very easy to check these formulas using the recursion formulas given above.

58. The n-th roots of 1 are $1, \varepsilon, \varepsilon^2, \dots, \varepsilon^{n-1}$ where

$$\varepsilon = \cos \frac{2\pi}{n} + i \sin \frac{2\pi}{n}.$$

The sum of k-th powers of these roots is $1 + \cdots + 1 = n$, if k is divisible by n, and it is 0 otherwise. The latter follows from the formula for the sum of a geometric sequence:

$$1 + \varepsilon^k + \varepsilon^{2k} + \cdots + \varepsilon^{(n-1)k} = \frac{\varepsilon^{kn} - 1}{\varepsilon^k - 1} = \frac{1-1}{\varepsilon^k - 1} = 0.$$

59. The curve $x = \cos 2t,\ y = \sin 3t$ is periodic with period 2π; so we can expect that this curve be closed. However, the graph (see Figure 59) does not look closed. The curve starts at the point $(1,0)$ $(t=0)$, then it goes up, to the point $\left(\frac{1}{2}, 1\right)$ $\left(t = \frac{\pi}{6}\right)$, and then, through the point $(0,0)$ $\left(t = \frac{\pi}{3}\right)$ to the point $(-1,-1)$ $\left(t = \frac{\pi}{2}\right)$. After that, the curve goes back along itself, because

$$\cos 2\left(\frac{\pi}{2} + \alpha\right) = \cos 2\left(\frac{\pi}{2} - \alpha\right) \text{ and } \sin 3\left(\frac{\pi}{2} + \alpha\right) = \sin 3\left(\frac{\pi}{2} - \alpha\right)$$

reaches $(0,0)$ at $t = \pi$, and then repeats the same path reflected in the x axis from $t = \pi$ to $t = 2\pi$.

The curve $x = t^3 - 3t, y = t^4 - 2t^2$ has two singularities (cusps) at $t = \pm 1$, since $x'(\pm 1) = y'(\pm 1) = 0$. On the graph, these cusps are located at the points $(\pm 2, -1)$ (see Figure 59).

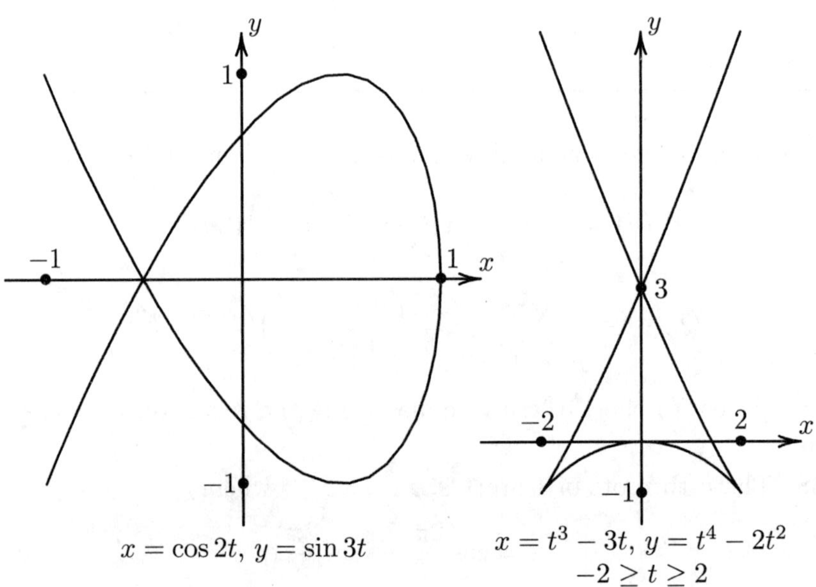

FIGURE 59. To Solution 59.

60. It is not hard to compute this integral precisely. For any $n \geq 2$, integration by parts gives

$$\begin{aligned}\int_0^{2\pi} \sin^n x\,dx &= -\sin^{n-1} x \cos x\Big|_0^{2\pi} - \int_0^{2\pi} (n-1)\sin^{n-2} x \cos x(-\cos x)\,dx \\ &= (n-1)\int_0^{2\pi} \sin^{n-2} x(1-\sin^2 x)\,dx \\ &= (n-1)\int_0^{2\pi} \sin^{n-2} x\,dx - (n-1)\int_0^{\pi} \sin^n x\,dx,\end{aligned}$$

so that

$$n\int_0^{2\pi} \sin^n x\,dx = (n-1)\int_0^{2\pi} \sin^{n-2} x\,dx$$

or

$$\int_0^{2\pi} \sin^n x\,dx = \frac{n-1}{n}\int_0^{2\pi} \sin^{n-2} x\,dx$$

(actually, if n is odd, then this equality becomes $0 = 0$, but this is not important to us now). Since $\int_0^{2\pi} \sin^0 x\,dx = 2\pi$, we have

$$\int_0^{2\pi} \sin^{100} x\,dx = \frac{99}{100}\cdot\frac{97}{98}\cdot\frac{95}{96}\cdot\ldots\cdot\frac{1}{2}\cdot 2\pi = \frac{100!}{2^{100}(50!)^2}\cdot 2\pi.$$

It is not hard to compute this number explicitly using a pocket calculator. The result is

$$0.079589 \cdot 2\pi \approx 0.500072.$$

Or one can use Stirling's approximation of factorials, $n! \approx \sqrt{2\pi n}\left(\frac{n}{e}\right)^n$; then our expression is approximated by

$$\frac{\sqrt{2\pi 100}}{(\sqrt{2\pi 50})^2}\frac{1}{2^{100}}\frac{100^{100}}{e^{100}}\frac{e^{100}}{50^{100}}\cdot 2\pi = \frac{\sqrt{2\pi}}{5} \approx 0.501236.$$

Both computations show that the approximation

$$\int_0^{2\pi} \sin^{100} x\,dx \approx 0.5$$

gives an error much less than the problem requested.

61. The logarithmic differentiation gives

$$(x^x)' = x^x(1+\ln x).$$

Hence, by the Fundamental Theorem of Calculus,

$$\int_1^{10} x^x(1+\ln x)\,dx = x^x]_1^{10} = 10^{10} - 1^1.$$

It is clear, however, that the function $y = x^x$, $1 \geq x \geq 10$, is concentrated in a proximity of $x = 10$ (see the graph in Figure 60).

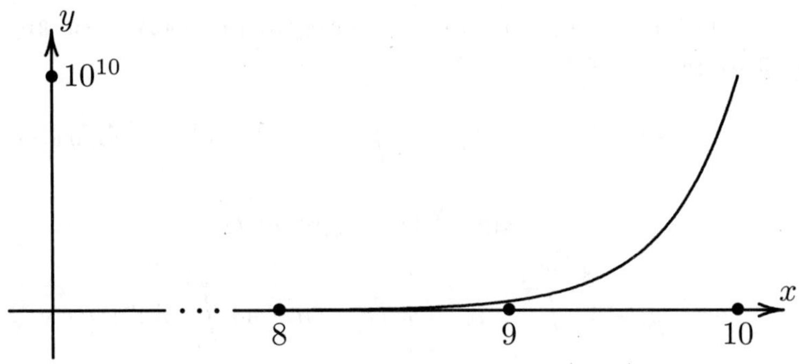

FIGURE 60. To Solution 61.

Hence, if we assume $\int_1^{10} x^x(1+\ln x)\,dx \approx \int_1^{10} x^x(1+\ln 10)\,dx$, then the error will be relatively small (we omit a precise estimate, but an easy computation shows that the relative error will be less than 1%, not 10%). If we assume this, we will get

$$\int_1^{10} x^x\,dx \approx \frac{10^{10}}{1+\ln 10} \approx 3,027,931,065.6.$$

Remark. A computer computation of the given integral gives

$$\int_1^{10} x^x\,dx \approx 3,007,764,122.4.$$

Thus, the approximation $\int_1^{10} x^x\,dx \approx 3\cdot 10^9$ (three billion) is much better, than the problem requested.

62. Draw big circles which contain the sides of our triangle. Let T be our triangle. We denote by S_α the spherical sector bounded by two great semicircles obtained by continuation of the sides of the triangle forming the angle α (so $T \subset S_\alpha$), and define in a similar way spherical sectors S_β and S_γ. See Figure 61. We will use the same notations $T, S_\alpha, S_\beta, S_\gamma$ for the areas of these four domains. Obviously, S_α is $\dfrac{\alpha}{2\pi}$ of the whole sphere, so $S\alpha = 2\alpha$; similarly, $S_\beta = 2\beta$ and $S_\gamma = 2\gamma$.

Consider the three hemispheres bounded by the three great circles which contain T. Their common area is $3\cdot 2\pi = 6\pi$, and they cover the whole sphere minus the triangle T' antipodal to T; thus, they cover the area $4\pi - T$. However, they overlap: each of the differences $S_\alpha - T, S_\beta - T$, and $S_\gamma - T$ is covered twice, and T is covered thrice. Hence, to obtain the area covered by the three hemispheres, we need to subtract from 6π each of $S_\alpha - T$, $S_\beta - T$, $S_\gamma - T$, and also $2T$. This yields the relation

$$4\pi - T = 6\pi - (2\alpha - T) - (2\beta - T) - (2\gamma - T) - 2T,$$

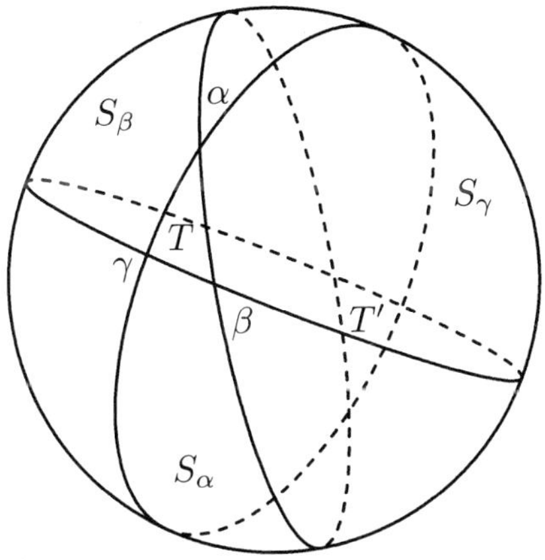

FIGURE 61. To Solution 62.

which gives, after cancellations, the equality

$$T = \alpha + \beta + \gamma - \pi.$$

63. Below are drawings of hypocycloids for four values of r.

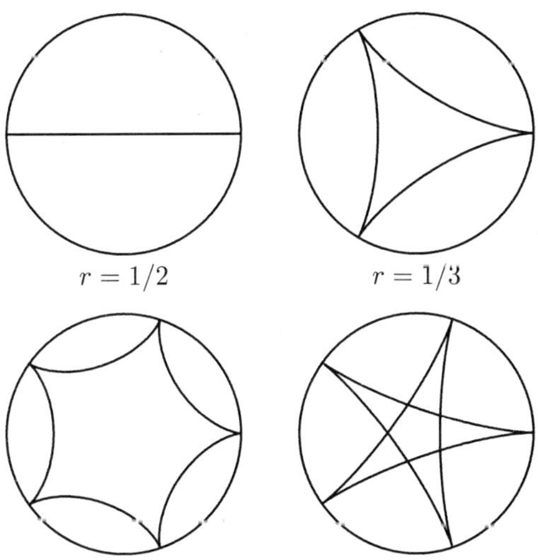

FIGURE 62. To Solution 63.

64. Let N be the number of days in a year. For different values of the number n of students in class, let us find the probability $p(n)$ of *no coincidences* in birthdays. If $n = 1$, then the probability is 1. If we add one student, then "no coincidences" means that the birthday of the newcomer is not the same as that of the first one; thus $p(2) = \dfrac{N-1}{N}$. For the third student, we get a new condition, independent of the previous one: the birthday of the new one should not fall on the birthdays of the previous two. Thus, $p(3) = \frac{N-1}{N}\frac{N-2}{N}$. And so on:

$$p(n) = \frac{N-1}{N}\frac{N-2}{N}\cdots\frac{N-(n-1)}{N} = \left(1-\frac{1}{N}\right)\cdots\left(1-\frac{n-1}{N}\right).$$

From this:

$$\ln p(n) = \ln\left(1-\frac{1}{N}\right) + \ln\left(1-\frac{2}{N}\right) + \cdots + \ln\left(1-\frac{n-1}{N}\right).$$

Using the approximation $\ln(1-t) = -t$ (for t small enough), we find that

$$\ln p(n) \approx -\frac{1+2+\cdots+(n-1)}{N} = -\frac{n(n-1)}{2N}.$$

We want to find the value n_0 of n for which the last expression is close to $\ln\frac{1}{2}$, that is, $n_0(n_0-1) \approx 2N\ln 2$. When $N = 365$, the right hand side of the last formula is $\approx 506 = 23 \cdot 22$, so we can take $n_0 = 23$. To make this approximate calculation more convincing, let us observe some calculator values of the probability $1-p(n)$ of existing of a pair of students with the same birthday:

$n = 10$	11.7%;	$n = 30$	70.6%;
$n = 15$	25.3%;	$n = 40$	89.1%;
$n = 23$	50.7%;	$n = 50$	97.4%.

66. The vertices of the triangle KLM of the minimal perimeter are the bases of the altitudes AL, BM, and CK. To prove this, consider an arbitrary triangle KLM inscribed into the triangle ABC and then reflect the triangle ABC first in the side BC, and in the image BA' of the side BA. These reflections map the side LK (of the inscribed triangle) onto LK' and the side KM onto $K'M'$. (See the left side of Figure 63.)

The perimeter of the triangle KLM is equal to the length of the polygonal line $MLK'M'$, and it is clear that this perimeter is not minimal, if this polygonal line is not straight. If it is straight, then $\angle CLM = \angle BLK' = \angle BLK$. Similarly, we must have $\angle LK'B = \angle M'K'A'$ which is the same as $\angle LKB = \angle MKA$. Also, the equality $\angle AMK = \angle CML$ must be true, because the polygonal line $LK'M'L'$ (obtained by one more reflection of the triangle) must be also straight. All these equalities of angles hold if K, L, and M are bases of altitudes, as shown in the drawing on the right.

If the triangle is obtuse, then this construction does not go through, since two of the three altitudes lie outside of the triangle. In this case the inscribed triangle of the minimal perimeter is the degenerate triangle AKA where A is the vertex of the obtuse angle and AK is the altitude from this vertex.

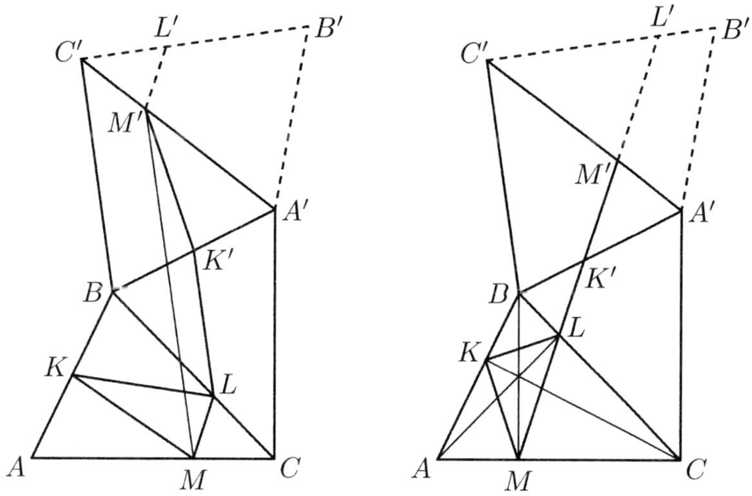

FIGURE 63. To Solution 66.

67. Let $\rho = \sqrt{X^2+Y^2+Z^2}$ (that is, the value of our function at the point (X,Y,Z)). Then the average of the function $\frac{1}{r}$ with respect to the sphere of radius R centered at (X,Y,Z) is equal to $\frac{1}{\rho}$ if $\rho \geq R$ and is equal to $\frac{1}{R}$ if $\rho \leq R$. Let us prove this.

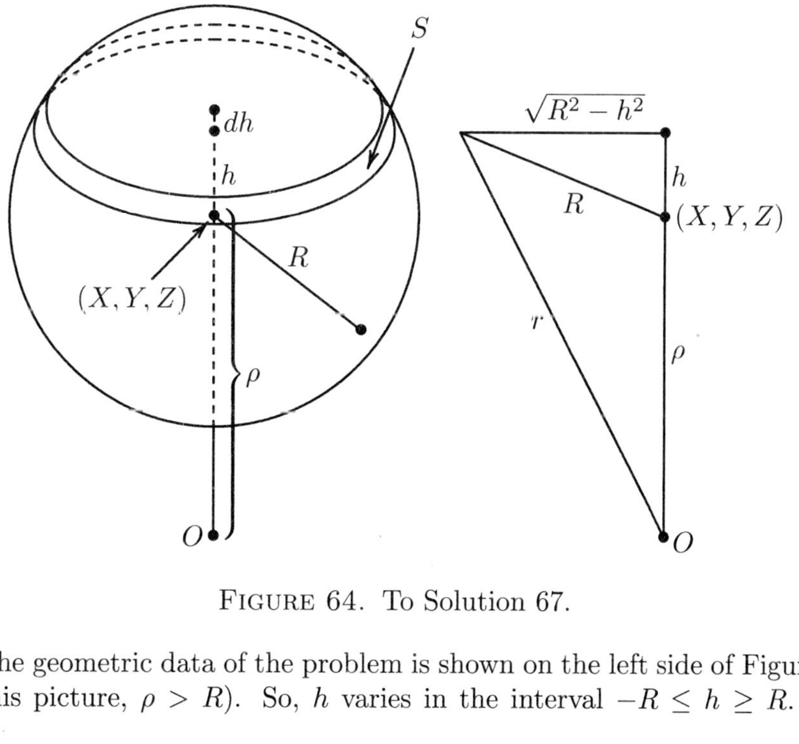

FIGURE 64. To Solution 67.

The geometric data of the problem is shown on the left side of Figure 64 (in this picture, $\rho > R$). So, h varies in the interval $-R \leq h \geq R$. Let

S be the spherical belt between planes perpendicular to the ray from O to (X, Y, Z) at the levels h and $h + dh$. It is known (see Solution to Problem 28) that the area of S is the same as the area of its projection onto the cylinder circumscribed about the sphere, that is, it is equal to $2\pi R \cdot dh$. Our function is nearly constant within S (we assume dh very small); the right side of Figure 64 shows that

$$r^2 = (\rho + h)^2 + (R^2 - h^2) = \rho^2 + 2\rho h + R^2.$$

To find the average value of the function with respect to the sphere, we need to find the integral of this function over the sphere and divide it over the area $4\pi R^2$ of the sphere. Thus, our average is

$$\frac{1}{4\pi R^2} \int_{-R}^{R} \frac{2\pi R\, dh}{\sqrt{\rho^2 + 2\rho h + R^2}} = \frac{1}{2R} \int_{-R}^{R} \frac{dh}{\sqrt{\rho^2 + 2\rho h + R^2}}.$$

Make a substitution $\rho^2 + 2\rho h + R^2 = u$. Then $du = 2\rho dh$, and if $h = \pm R$, then $u = \rho^2 \pm 2\rho R + R^2 = (\rho \pm R)^2$. We continue computing the average:

$$= \frac{1}{2R} \int_{(\rho-R)^2}^{(\rho+R)^2} \frac{du}{2\rho\sqrt{u}} = \frac{1}{4R\rho} \cdot 2\sqrt{u}\Big]_{(\rho-R)^2}^{(\rho+R)^2}$$

$$= \frac{1}{2R\rho}((\rho + R) - |\rho - R|) = \begin{cases} \dfrac{1}{\rho}, & \text{if } \rho \geq R, \\ \dfrac{1}{R}, & \text{if } \rho \leq R, \end{cases}$$

as was stated.

Remark. If $\rho = R$, the integral becomes improper (the integrand is ∞ for $h = -R$), but this does not affect its value $\dfrac{1}{\rho} = \dfrac{1}{R}$.

68. (Here and below, we use the notation log for $\log_{10}$.) $10 \cdot \log 2 = \log 2^{10} = \log 1024 = \log 1000 + \log 1.024 = 3 + \log 1.024$. It is not hard to check that $\log 1.024$ is very close to 0.01 ($\sqrt[3]{10} \approx 2.1544$ and, taking 5 times the square root of this number, we see that $\sqrt[96]{10} \approx 1.0243$). Hence, $10 \log 2 \approx 3.01$ and $\log 2 \approx 0.301$.

Actually, the difference between $\log 1.024$ and 0.01 is much less than 0.001, so the approximation of $\log 2$ by 0.301 is better than just three decimal places. The calculator value of $\log 2$ is 0.30103.

69. (It is reasonable to have in mind the last remark in the previous solution.)

$\log 4 = 2 \log 2 \approx 0.602$;
$\log 8 = 3 \log 2 \approx 0.903$;
$\log 5 = 1 - \log 2 \approx 0.699$;
$\log 50 = 1 + \log 5 \approx 1.699$;
$\log 32 = 5 \log 2 \approx 1.505$;
$\log 128 = 7 \log 2 \approx 2.107$;
$\log 125 = 3 \log 5 \approx 2.097$;
$\log 64 = 6 \log 2 \approx 1.808$.

70. A rough estimate would be $\log 7 = \frac{1}{2}\log 50 \approx \frac{1}{2} \cdot 1.699$. We could obtain a better approximation, if we notice that $\frac{50}{49} \approx 1.02$:

$$\log 50 \approx \log 49 + \log 1.02 \approx 2\log 7 + 0.01,$$

so

$$\log 7 \approx \frac{1}{2}(1.699 - 0.01) \approx 0.845.$$

71. Roughly, $6\log 2 = \log 64 \approx \log 63 - \log 7 + \log 9$, from which $\log 9 \approx 6\log 2 - \log 7 \approx 6 \cdot 0.301 - 0.845 = 0.961$. To get a better approximation, we notice that the function $\log(1+x)$ is 0 at 0 and, as any smooth function, is almost linear for small values of x. Since we know that $\log 1.024 \approx 0.01$ and $\frac{64}{63} \approx 1.016$, we can conclude that $\log 6463 \approx \frac{0.016}{0.024}\log 1.024 \approx 0.007$. Hence,

$$6\log 2 = \log 63 + \log\frac{64}{63} \approx \log 7 + \log 9 + 0.007,$$
$$\log 9 \approx 6 \cdot 0.301 - 0.845 - 0.007 = 0.954.$$

Furthermore,

$$\log 3 = \log 9/2 \approx 0.477; \qquad \log 6 = \log 2 + \log 3 \approx 0.778;$$
$$\log 27 = 3\log 3 \approx 1.431; \qquad \log 12 = 2\log 2 + \log 3 \approx 1.079.$$

72. $\log 1.024 \approx 0.01$, while $\ln 1.024 \approx 0.024$. Therefore,

$$\ln 10 \approx \frac{0.024}{0.01} = 2.4, \text{ and } \log e = \frac{1}{\ln}10 \approx \frac{1}{2.4} \approx 0.42.$$

Remark. As usual, multiple approximations lead to significant errors. The calculator values of $\ln 10$ and $\log e$ are slightly different: $\ln 10 \approx 2.3026$ and $\log e \approx 0.4343$.

73, 74. We begin with a computer result. For a digit $n = 1, 2, \ldots, 9$, we denote by $d_2(n)$ the number of powers of 2, among $2^1, 2^2, 2^3, \ldots, 2^{10000}$, with the first digit n, by $d_3(n)$ the similar number for powers of 3, and put $L(n) = [10000 \cdot (\log(n+1) - \log(n))]$. Here are the values of $d_2(n), d_3(n)$ and $L(n)$:

n	$d_2(n)$	$d_3(n)$	$L(n)$
1	3011	3007	3010
2	1760	1764	1761
3	1250	1247	1249
4	969	968	969
5	791	792	792
6	670	669	669
7	580	582	580
8	511	513	512
9	458	458	458

This table speaks for itself. It strongly suggests that if $d_2(n, N)$ is the number of powers of 2, among $2^1, 2^2, \ldots, 2^N$ with the first digit n, then

$$\lim_{N\to\infty} \frac{d_2(n, N)}{N} = \log(n+1) - \log n,$$

and the same is true for powers of 3. Actually, as we explain below, this is true for powers of any positive real number a such that $\log a$ is irrational.

Lemma. *Let α be a positive irrational number, and let γ_m be the fractional part of the number $m\alpha$ (that is, $\gamma_m = m\alpha - [m\alpha]$). Choose any interval $[c, d] \subset [0, 1]$, and denote by $F(n)$ the number of elements of the set $\{\gamma_1, \gamma_2, \ldots, \gamma_n\}$ which are contained in $[c, d]$. Then*

$$\lim_{n\to\infty} \frac{F(n)}{n} = d - c.$$

Proof. Let $[c', d'] \subset [0, 1]$ be the interval obtained from $[c, d]$ by shifting right by $k\alpha$ (within the whole number line) and then shifting left by some integer (in other words, $c + k\alpha - c'$ and $d + k\alpha - d'$ are equal integers; in particular, $d' - c' = d - c$). Let γ_i' and $F'(n)$ be defined for $[c', d']$ in the same way as γ_i, and let $F(n)$ be defined for $[c, d]$. Then $\gamma_m' = \gamma_{m+k}$, which shows that, for any n, $|F'(n) - F(n)| \leq k$. Thus, for n large, the ratios $\dfrac{F(n)}{n}$ and $\dfrac{F'(n)}{n}$ are very close to each other.

Next we prove that for n large the ratio $\dfrac{F(n)}{n}$ does not exceed twice the length of the interval $[c, d]$. Suppose that $\dfrac{1}{r} \leq d - c < \dfrac{1}{r-1}$ where r is an integer and $r \geq 2$. In the case $r = 2$ we have nothing to prove (since twice the length is at least 1); so we may assume that $r \geq 3$. We can find mutually disjoint intervals, $[c_1, d_1], \ldots, [c_{r-1}, d_{r-1}]$, each of the form $[c', d']$ for some k. For an n really big, the ratios $\dfrac{F_i(n)}{n}$ (calculated for intervals $[c_i, d_i]$) are almost the same, and since their sum does not exceed 1 (because $F_1(n) + \cdots + F_{r-1}(n) \leq n$), each of them, for big n, will be less than any number greater $\dfrac{1}{r-1}$, in particular, some number less than $\dfrac{2}{r} \leq 2(d-c)$; this is what we wanted to prove.

Therefore, if we slightly change c and d, then for n large the ratios $\dfrac{F(n)}{n}$ will stay almost unchanged. In particular, if two intervals have the same length, then these ratios for them are almost the same for n large.

From this we easily deduce our statement. For the intervals $\left[0, \dfrac{1}{r}\right]$, $\left[\dfrac{1}{r}, \dfrac{2}{r}\right], \ldots, \left[\dfrac{r-1}{r}, 1\right]$ the ratios $\dfrac{F(n)}{n}$ are almost the same for n large, and their sum is 1. Thus, $\lim\limits_{n\to\infty} \dfrac{F(n)}{n} = \dfrac{1}{r} =$ length, and this is true for any interval of the length $\frac{1}{r}$. Consequently, it is true for any interval of rational length, and hence, by the remark above, for any interval at all. $\square$

The lemma implies our statement. Let γ_n be the fractional part of $n \log a$. The first digit of a^n is 1, if γ_n lies in the interval $[0, \log 2]$; it is 2, if γ_n lies in $[\log 2, \log 3]$; and so on. Our limit relation follows.

Remark. This distribution of the first digits can be observed not only for the sequence of powers of a fixed real number, but, in some sense, for any naturally defined sequence of numbers. (The reader who likes experimenting, can take, for example, the list of all cities in California. The populations will have first digits distributed in the same way.) This phenomenon is called "Benford's Law", after F. Benford who described it in 1938. (F. Benford was, in turn, inspired by observations made by S. Newcomb in 1881.) A rigorous mathematical explanation of Benford's Law is still missing. A good reference for this is the article "Benford's Law strikes back: no simple explanation in sight for mathematical gem" by A. Berge and T. P. Hill [**18***].

75. The domains $U, g(U), g^2(U), \ldots$ cannot be all disjoint, since the area of M is finite, and the area of U is positive. Hence, some intersection $g^m(U) \cap g^n(U)$ with $m > n$ is non-empty. So, there exist $x, y \in U$ such that $g^m(x) = g^n(y)$. Applying g backward n times, we get $g^{m-n}(x) = y \in U$, and we can take $T = m - n$.

76. First, notice that this density property does not depend on the choice of $x \in M$. Indeed, if $x = (\xi, \eta) \bmod 2\pi$ and $x' = (\xi', \eta') \bmod 2\pi$, then the set $\{x', gx', g^2x', \ldots\}$ can be obtained from $\{x, gx, g^2x, \ldots\}$ by the transformation $(\alpha, \beta) \mapsto (\alpha - \xi + \xi', \beta - \eta + \eta') \bmod 2\pi$ which, obviously, would not affect the density. Second, we remark that to prove our statement we need to know that $1, \sqrt{2}$, and π are not comeasurable, that is, for no non-zero triple of integers p, q, r the number $p + q\sqrt{2} + r\pi$ is zero. The last fact follows from the statements that $\sqrt{2}$ is irrational and π is transcendental; these statements are broadly known, but the proof of the second one is not elementary, and we will not give it here.

Fix a small $\varepsilon > 0$, and let $x = (\pi, \pi)$. Cover the square $\{0 \leq x \leq 2\pi, 0 \leq y \leq 2\pi\} \subset \mathbb{R}^2$ by a finite family of disks $d_1, \ldots, d_N$ of diameters $< \varepsilon$. For every n, the point g^n belongs to some d_r modulo 2π. Since the number of disks is finite, it is true that for some $k, \ell > k$, the points $g^k x, g^\ell x$ belong (modulo 2π) to the same d_r. Hence, $g^n x$, where $n = \ell - k$, lies in the disk of radius $\varepsilon/2$ centered at x. We will prove the following: the set $\{x, g^n x, g^{2n} x, g^{3n} x, \ldots\}$ is ε-dense in the torus, that is, for every point of the torus, there exists a point of our set at the distance $< \varepsilon$ from this point. Since ε is arbitrary, this shows that the set $S = \{x, gx, g^2x, \ldots\}$ is dense in the torus.

Modulo 2π, $g^n x$ is $y = (\pi + n + 2M\pi, \pi + n\sqrt{2} + 2N\pi)$ where M, N are integers and the distance δ from y to x is less than ε. Let L be the line in the plane passing through the points x and y. If we take on L the points at the distance $0, \delta, 2\delta, \ldots$ from x, then modulo 2π this will be our set S. Thus, S may be regarded as an $\varepsilon/2$-dense subset of the line L.

The line L intersects the horizontal line $y = 3\pi$ at some point $(\pi + \lambda, 3\pi)$ where $\lambda = 2\pi \dfrac{n + 2M\pi}{n\sqrt{2} + 2N\pi}$. It is important for us that $\dfrac{\lambda}{\pi}$ is not rational. But indeed, if $\dfrac{\lambda}{\pi} = \dfrac{p}{q}$, then

$$\frac{p}{q} = \frac{2n + 4M\pi}{n\sqrt{2} + 2N\pi},$$

so

$$p(n\sqrt{2} + 2N\pi) = q(2n + 4M\pi),$$

which obviously contradicts to the above-mentioned fact that 1, $\sqrt{2}$, and π are not comeasurable.

Notice now in Figure 65 that the horizontal shift of the line L by λ units produces the same line L' as the vertical shift of L by -2π units:

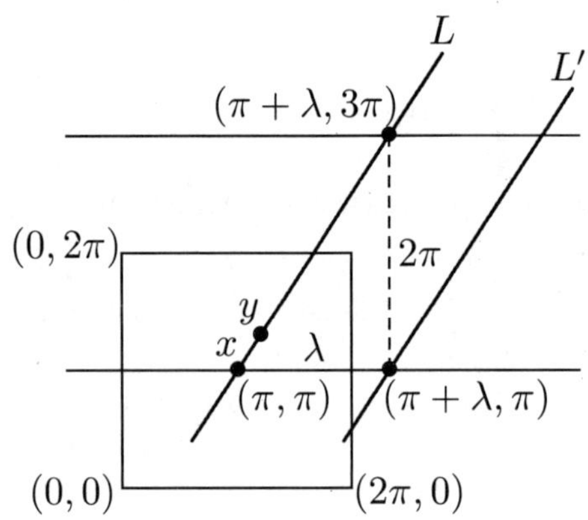

FIGURE 65. To Solution 76.

Obviously, from the point of view of the torus, L and L' are the same line (because L' is obtained from L by a vertical shift by -2π). Consequently, for arbitrary integers p, q, a horizontal shift of the line L by $p\lambda + 2q\pi$ does not change it as a subset of the torus. In particular, the set S can be regarded as an $\varepsilon/2$-dense subset of any such line. (Our proof will show, actually, that this set is dense, but we will not need that.)

Next, let us prove that the set of numbers of the form $p\lambda + 2q\pi$ is $\varepsilon/2$-dense in the real line. It is important that we already know that all these points are pairwise different (since λ/π is irrational). The proof is similar to first step of the current proof. Obviously, there are infinitely many points of our set in the interval $[0, 2\pi]$ (for every p there exists a q such that $p\lambda + 2q\pi \in [0, 2\pi]$). Cover the interval $[0, \pi]$ by intervals $i_1, \ldots, i_N$ of lengths $< \varepsilon$. Since the number of intervals is finite, it is possible to find two different points of our set, $p_1\lambda + 2q_1\pi$ and $p_2\lambda + 2q_2\pi$ which belong to the same interval; we may assume that the second of these numbers is greater, than the first one. Then the number $p\lambda + 2q\pi$ where $p = p_2 - p_1$ and $q = q_2 - q_1$ lies in the interval $[0, \varepsilon]$, and the points $mp\lambda + 2mq\pi$ with $0 < m < 2\pi/\varepsilon$ form an $\varepsilon/2$-dense subset of $[0, 2\pi]$.

Now we can finish our proof. For every point (α, β) of the torus, there is a line (in the plane) at the distance less that $\varepsilon/2$ from (α, β) which contains S as an $\varepsilon/2$-dense subset. The point of this subset closest to (α, β) is at the distance $< \varepsilon$ from (α, β). This is all we need.

77. Every point of the form $(r\pi, s\pi)$ with rational r and s is periodic. Indeed, let q be a common denominator of r and s, so our point (α, β) has the form $\left(\frac{a}{q}\pi, \frac{b}{q}\pi\right)$ with non-negative integer $a < 2q$ and $b < 2q$. Then, all points $g(\alpha, \beta), g^2(\alpha, \beta), g^3(\alpha, \beta), \ldots$ have the same form (maybe, with different pairs a, b). But there are finitely many (at most $4q^2$) such points. Hence there must be an equality $g^m(\alpha, \beta) = g^n(\alpha, \beta)$ for some $m > n$. Our transformation g is invertible $g^{-1}(\alpha, \beta) = (\alpha - \beta, -\alpha + 2\beta)$. Apply n times g^{-1} to the equality $g^m(\alpha, \beta) = g^n(\alpha, \beta)$, and we will get $g^{m-n}(\alpha, \beta) = (\alpha, \beta)$. It is obvious that the set of points of this form is dense in the torus.

79. The solution is similar to our solution of Problem 74.

Bibliography

[1] V. B. Alekseev. *Abel's Theorem in Problems and Solutions*, Springer-Verlag, 2004.
[2] I. V. Arnold. *Number Theory* (Russian), Uchpedgiz, Moscow, 1939.
[3] V. I. Arnold. *Modes and Quasimodes*, Funct. Anal. Appl., 1972, **6**, 94–101.
[4] V. I. Arnold. *Mathematical Methods of Classical Mechanics*, Springer-Verlag, 1989.
[5] V. I. Arnold. *A-graded algebras and continued fractions*, Comm. Pure Appl. Math. 1989, **42**, 993–1000.
[6] V. I. Arnold. *Higher dimensional continued fractions*, Regular and Chaotic Dynamics, 1998, **3**, No. 3, 10–17.
[7] V. I. Arnold. *Weak asymptotics of the numbers of solutions of Diophantine equations.* Funct. Anal. Appl., 1999, **33**, 292–293.
[8] V. I. Arnold. *Polymathematics: Is mathematics a single science or a set of arts?*, in the book "Mathematics: Frontiers and Perspectives", AMS, 2000, 403–416.
[9] V. I. Arnold. *Relatives of he quotient of the complex projective plane by the complex conjugation*, Proc. Steklov Math. Inst., **224**, 46–56.
[10] V. I. Arnold. *The Lagrangian Grassmannnian of Quaternion Hypersymplectic Space*, Funct. Anal. Appl., 2001, **35**, 61–63.
[11] V. I. Arnold. *Pseudoquaternion geometry*, Funct. Aanal. Appl., 2002, **36**, 1–12.
[12] V. I. Arnold. *The Fermat-Euler dynamical system and the statistics of the arithmetic of geometric progressions*, Funct. Anal. Appl., 2003 **37** 1–15.
[13] V. I. Arnold. *Arithmetics of binary quadratic forms, symmetry of their continued fractions and geometry of their de Sitter world*, Bull. Brasil Math. Soc. 2003, **3**, 1–41.
[14] V. I. Arnold. *Topology and statistics of formulas of arithmetic*, Russian Math. Surveys, 2003, **58**, 637–664.
[15] V. I. Arnold. *The topology of algebra: combinatorics of squaring*, Funct. Anal. Appl., 2003, **37**, 177–190.
[16] V. I. Arnold. *Ergodic and arithmetical properties of geometrical progression's dynamics and of its orbits,* Mosc. Math. J., 2005, **5**, 5–22.
[17] V. O. Bugayenko. *Pell Equations* (Russian). Moscow, MCNMO, 2001. (Library "Mathematical Education", Vol. 13).
[18*] A. Berge and T. P. Hill. *Benford's Law strikes back: no simple explanation in sight for mathematical gem*, Mathematical Intellegencer, 2011, **33.1**, 85–91.
[19] G. L. Dirichlet, *Über die Bestimmung der mittleren Werte in der Zahlentheorie*, Abh. Akad. Wiss. Berlin (1849), 78–81.
[20] L. Euler. *Commentationes arithmeticae collectae*, vol. 1–2, St. Petersburg, 1849.
[21] R. Fauré. *Transformations Conformes en Mécanique Ondulatoire*, C. R. Acad. Sci. Paris, 237 (1953), 603–605.
[22] P. Fermat. *Œvres de Fermat*, vol. I–IV, Gauthier-Villars, Paris, 1891–1912.
[23*] D. Fuchs and S. Tabachnikov. *Mathematical Omnibus*, Amer. Math. Soc., 2007.

[24] H. Gylden. *Quelques remarques rélativement à la représentation des nombres irrationels par des fraction continues.* C. R. Acad. Sci. Paris, 1888, **107**, 1584–1587.

[25] C. G. J. Jacobi. *Canon Arithmeticus*, Berolini, 1839.

[26] E. Kasner. *Differential-Geometric Aspects of Dynamics*, Amer. Math. Soc., 1913.

[27] A. Ya. Khinchin. *Continued Fractions*, Dover Publ., NY, 1997.

[28] E. Korkina. *La périodicité des fractions continues multidimensionelles.*, C. R. Acad. Sci. Paris. Ser. I. 1994, **319**, 777–780.

[29] E. Korkina. *Two-dimensional continued fractions. The simplest examples*, Proc. Steklov Math. Inst., 1995, **209**, 124–144.

[30] T. Needham. *Newton and Transmutation of Force*, Amer. Math. Monthly, 1993, **100**, 119–137.

[31] L. S.P olak. *Variational Principles of Mechanics, their Developments and Application to Physics* (Russian). Fizmatlit, Moscow, 1960.

[32] B. Sturmfels. *Gröbner bases and convex polytopes*, Providence, R. I.: AMS, 1996. (University Lecture Series. vol. 8), 85–98.

[33] H. Tsushiyashi. *Higher-dimensional analogues of periodic continued fractions and cusp singularities*, Tohoku Math. J. 1983, **35**, 607–639.

[34] B. A. Venkov. *Elementary Number Theory*, Wolters-Noordfof, Groninhen, 1970.

[35] A. Wiman. *Über eine Wahrscheinlichkeits auflage bei Kettenbruchentwicklungen*, Akad. Föhr. Stockholm, 1900, **57**, 589–841.

[36] V. Arnold, M. Kontsevich, and A. Zorich, editors, *Pseudoperiodic Topology*, AMS Translations. Ser. 2. vol. 197 (Advances in Mathematical Sciences. vol. 46), American Mathematical Society, Providence, RI, 1999, 9–27.

[37] *Modern Mathematics* (Russian), Moscow–Dubna 2002, MTsNMO, 2002.

图字：01-2023-0918 号

讲义与问题：
给青年数学家的礼物
Jiangyi yu Wenti :
Gei Qingnian Shuxuejia de Liwu

图书在版编目 (CIP) 数据

讲义与问题：给青年数学家的礼物：影印版 = Lectures and Problems: A Gift to Young Mathematicians : 英文 / (俄罗斯) V. I. 阿诺尔德 (V. I. Arnold) 著；(美) 德米特里 · 福克斯 (Dmitry Fuchs)，(美) 马克 · 索尔 (Mark Saul) 译. — 北京：高等教育出版社，2024.2
ISBN 978-7-04-061160-1
Ⅰ. ①讲… Ⅱ. ① V… ②德… ③马… Ⅲ ①数学—青少年读物—英文 Ⅳ. ① O1-49
中国国家版本馆 CIP 数据核字 (2023) 第 173142 号

策划编辑 李　鹏　　责任编辑 李　鹏
封面设计 张申申　　责任印制 沈心怡

出版发行 高等教育出版社
社址 北京市西城区德外大街 4 号
邮政编码 100120
购书热线 010-58581118
咨询电话 400-810-0598
网址 http://www.hep.edu.cn
http://www.hep.com.cn
网上订购 http://www.hepmall.com.cn
http://www.hepmall.com
http://www.hepmall.cn
印刷 涿州市星河印刷有限公司

开本 787mm × 1092mm 1/16
印张 12.25
字数 290 千字
版次 2024 年 2 月第 1 版
印次 2024 年 2 月第 1 次印刷
定价 99.00 元

美国数学会经典影印系列

1 **Lars V. Ahlfors**, Lectures on Quasiconformal Mappings, Second Edition 9787040470109

2 **Dmitri Burago**, **Yuri Burago**, **Sergei Ivanov**, A Course in Metric Geometry 9787040469080

3 **Tobias Holck Colding**, **William P. Minicozzi II**, A Course in Minimal Surfaces 9787040469110

4 **Javier Duoandikoetxea**, Fourier Analysis 9787040469011

5 **John P. D'Angelo**, An Introduction to Complex Analysis and Geometry 9787040469981

6 **Y. Eliashberg**, **N. Mishachev**, Introduction to the *h*-Principle 9787040469028

7 **Lawrence C. Evans**, Partial Differential Equations, Second Edition 9787040469356

8 **Robert E. Greene**, **Steven G. Krantz**, Function Theory of One Complex Variable, Third Edition 9787040469073

9 **Thomas A. Ivey**, **J. M. Landsberg**, Cartan for Beginners: Differential Geometry via Moving Frames and Exterior Differential Systems 9787040469172

10 **Jens Carsten Jantzen**, Representations of Algebraic Groups, Second Edition 9787040470086

11 **A. A. Kirillov**, Lectures on the Orbit Method 9787040469103

12 **Jean-Marie De Koninck**, **Armel Mercier**, 1001 Problems in Classical Number Theory 9787040469998

13 **Peter D. Lax**, **Lawrence Zalcman**, Complex Proofs of Real Theorems 9787040470000

14 **David A. Levin**, **Yuval Peres**, **Elizabeth L. Wilmer**, Markov Chains and Mixing Times 9787040469943

15 **Dusa McDuff**, **Dietmar Salamon**, *J*-holomorphic Curves and Symplectic Topology 9787040469936

16 **John von Neumann**, Invariant Measures 9787040469974

17 **R. Clark Robinson**, An Introduction to Dynamical Systems: Continuous and Discrete, Second Edition 9787040470093

18 **Terence Tao**, An Epsilon of Room, I: Real Analysis: pages from year three of a mathematical blog 9787040469004

19 **Terence Tao**, An Epsilon of Room, II: pages from year three of a mathematical blog 9787040468991

20 **Terence Tao**, An Introduction to Measure Theory 9787040469059

21 **Terence Tao**, Higher Order Fourier Analysis 9787040469097

22 **Terence Tao**, Poincaré's Legacies, Part I: pages from year two of a mathematical blog 9787040469950

23 **Terence Tao**, Poincaré's Legacies, Part II: pages from year two of a mathematical blog 9787040469967

24 **Cédric Villani**, Topics in Optimal Transportation 9787040469219

25 **R. J. Williams**, Introduction to the Mathematics of Finance 9787040469127

26 **T. Y. Lam**, Introduction to Quadratic Forms over Fields 9787040469196

27 **Jens Carsten Jantzen**, Lectures on Quantum Groups

28 **Henryk Iwaniec**, Topics in Classical Automorphic Forms

29 **Sigurdur Helgason**, Differential Geometry, Lie Groups, and Symmetric Spaces

30 **John B.Conway**, A Course in Operator Theory

31 **James E. Humphreys**, Representations of Semisimple Lie Algebras in the BGG Category *O*

32 **Nathanial P. Brown, Narutaka Ozawa**, C*-Algebras and Finite-Dimensional Approximations

33 **Hiraku Nakajima**, Lectures on Hilbert Schemes of Points on Surfaces

34 **S. P. Novikov, I. A. Taimanov, Translated by Dmitry Chibisov**, Modern Geometric Structures and Fields

35 **Luis Caffarelli, Sandro Salsa**, A Geometric Approach to Free Boundary Problems

36 **Paul H. Rabinowitz**, Minimax Methods in Critical Point Theory with Applications to Differential Equations

37 **Fan R. K. Chung**, Spectral Graph Theory

38 **Susan Montgomery**, Hopf Algebras and Their Actions on Rings

39 **C. T. C. Wall, Edited by A. A. Ranicki**, Surgery on Compact Manifolds, Second Edition

40 **Frank Sottile**, Real Solutions to Equations from Geometry

41 **Bernd Sturmfels**, Gröbner Bases and Convex Polytopes

42 **Terence Tao**, Nonlinear Dispersive Equations: Local and Global Analysis

43 **David A. Cox, John B. Little, Henry K. Schenck**, Toric Varieties

44 **Luca Capogna, Carlos E. Kenig, Loredana Lanzani**, Harmonic Measure: Geometric and Analytic Points of View

45 **Luis A. Caffarelli, Xavier Cabré**, Fully Nonlinear Elliptic Equations

46 **Teresa Crespo, Zbigniew Hajto,** Algebraic Groups and Differential Galois Theory

47 **Barbara Fantechi, Lothar Göttsche, Luc Illusie, Steven L. Kleiman, Nitin Nitsure, Angelo Vistoli,** Fundamental Algebraic Geometry: Grothendieck's FGA Explained

48 **Shinichi Mochizuki,** Foundations of p-adic Teichmüller Theory

49 **Manfred Leopold Einsiedler, David Alexandre Ellwood, Alex Eskin, Dmitry Kleinbock, Elon Lindenstrauss, Gregory Margulis, Stefano Marmi, Jean-Christophe Yoccoz,** Homogeneous Flows, Moduli Spaces and Arithmetic

50 **David A. Ellwood, Emma Previato,** Grassmannians, Moduli Spaces and Vector Bundles

51 **Jeffery McNeal, Mircea Mustaţă,** Analytic and Algebraic Geometry: Common Problems, Different Methods

52 **V. Kumar Murty,** Algebraic Curves and Cryptography

53 **James Arthur, James W. Cogdell, Steve Gelbart, David Goldberg, Dinakar Ramakrishnan, Jiu-Kang Yu,** On Certain L-Functions

54 **Rick Miranda,** Algebraic Curves and Riemann Surfaces — 9 787040 469066

55 **Hershel M. Farkas, Irwin Kra,** Theta Constants, Riemann Surfaces and the Modular Group — 9 787040 469042

56 **Fritz John,** Nonlinear Wave Equations, Formation of Singularities — 9 787040 517262

57 **Henryk Iwaniec, Emmanuel Kowalski,** Analytic Number Theory — 9 787040 517231

58 **Jan Malý, William P. Ziemer,** Fine Regularity of Solutions of Elliptic Partial Differential Equations — 9 787040 517248

59 **Jin Hong, Seok-Jin Kang,** Introduction to Quantum Groups and Crystal Bases — 9 787040 517255

60 **V. I. Arnold,** Topological Invariants of Plane Curves and Caustics — 9 787040 517057

61 **Dusa McDuff, Dietmar Salamon,** *J*-Holomorphic Curves and Quantum Cohomology — 9 787040 517040

62 **James Eells, Luc Lemaire,** Selected Topics in Harmonic Maps — 9 787040 517279

63 **Yuri I. Manin,** Frobenius Manifolds, Quantum Cohomology, and Moduli Spaces — 9 787040 517026

64 **Bernd Sturmfels,** Solving Systems of Polynomial Equations — 9 787040 517033

65 **Liviu I. Nicolaescu,** Notes on Seiberg-Witten Theory — 9 787040 469035

66 **I. G. Macdonald,** Symmetric Functions and Orthogonal Polynomials — 9 787040 535938

67 **Craig Huneke,** Tight Closure and Its Applications — 9 787040 535945

68 **Victor Guillemin, Viktor Ginzburg, Yael Karshon,** Moment Maps, Cobordisms, and Hamiltonian Group Actions — 9 787040 535952

69 **Gerald B. Folland,** Fourier Analysis and Its Applications — 9 787040 535969

70 **Luis Barreira, Yakov B. Pesin,** Lyapunov Exponents and Smooth Ergodic Theory — 9 787040 534962

71 **Gilles Pisier,** Factorization of Linear Operators and Geometry of Banach Spaces — 9 787040 535037

72 **Richard Elman, Nikita Karpenko, Alexander Merkurjev,** The Algebraic and Geometric Theory of Quadratic Forms — 9 787040 534955

73 **William Fulton,** Introduction to Intersection Theory in Algebraic Geometry — 9 787040 534870

74 **S. K. Lando,** Lectures on Generating Functions — 9 787040 535006

75 **Frank D. Grosshans, Gian-Carlo Rota, Joel A. Stein,** Invariant Theory and Superalgebras — 9 787040 534979

76 **Lawrence C. Evans,** Weak Convergence Methods for Nonlinear Partial Differential Equations — 9 787040 534993

77 **Harish-Chandra,** Admissible Invariant Distributions on Reductive *p*-adic Groups — 9 787040 535020

78 **Marc Levine,** Mixed Motives — 9 787040 535013

79 **Pavel I. Etingof, Igor B. Frenkel, Alexander A. Kirillov, Jr.,** Lectures on Representation Theory and Knizhnik-Zamolodchikov Equations — 9 787040 534986

80 **Nicholas M. Katz, Peter Sarnak,** Random Matrices, Frobenius Eigenvalues, and Monodromy — 9 787040 534948

81 **Kenneth R. Davidson,** C*-Algebras by Example — 9 787040 534924

82 **Max-Albert Knus, Alexander Merkurjev, Markus Rost, Jean-Pierre Tignol,** The Book of Involutions — 9 787040 534931

83 **Judy L. Walker,** Codes and Curves 9 787040 534917

84 **J. Amorós, M. Burger, K. Corlette, D. Kotschick, D. Toledo,** Fundamental Groups of Compact Kähler Manifolds 9 787040 536331

85 **V. I. Arnold,** Experimental Mathematics 9 787040 556261

86 **Leslie Jane Federer Vaaler, James W. Daniel,** Mathematical Interest Theory, Second Edition 9 787040 556278

87 **Paul R. Halmos,** Linear Algebra Problem Book 9 787040 556285

88 **Luis Barreira, Claudia Valls,** Ordinary Differential Equations: Qualitative Theory 9 787040 556315

89 **Terence Tao,** Topics in Random Matrix Theory 9 787040 556476

90 **Gerald Teschl,** Ordinary Differential Equations and Dynamical Systems 9 787040 556483

91 **Lawrence C. Evans,** An Introduction to Stochastic Differential Equations 9 787040 556490

92 **Gerald Teschl,** Mathematical Methods in Quantum Mechanics: With Applications to Schrödinger Operators, Second Edition 9 787040 556506

93 **András Vasy,** Partial Differential Equations: An Accessible Route through Theory and Applications 9 787040 569773

94 **Weinan E, Tiejun Li, Eric Vanden-Eijnden,** Applied Stochastic Analysis 9 787040 556391

95 **David R. Mazur,** Combinatorics: A Guided Tour 9 787040 556339

96 **David Bressoud,** A Radical Approach to Real Analysis 9 787040 556322

97 **H. Iwaniec,** Lectures on the Riemann Zeta Function 9 787040 556308

98 **Terence Tao,** Hilbert's Fifth Problem and Related Topics 9 787040 556292

99 **David Lannes,** The Water Waves Problem: Mathematical Analysis and Asymptotics 9 787040 556346

100 **Seán Dineen,** Probability Theory in Finance: A Mathematical Guide to the Black-Scholes Formula, Second Edition 9 787040 556353

101 **Steven G. Krantz,** Complex Analysis: The Geometric Viewpoint, Second Edition 9 787040 556360

102 **A. Borel, N. Wallach,** Continuous Cohomology, Discrete Subgroups, and Representations of Reductive Groups, Second Edition 9 787040 556377

103 **H. S. M. Coxeter,** Non-Euclidean Geometry, Sixth Edition 9 787040 556384

104 **Kurt O. Friedrichs**, Mathematical Methods of Electromagnetic Theory 9 787040 569810

105 **Jeffrey Rauch,** Hyperbolic Partial Differential Equations and Geometric Optics 9 787040 569803

106 **Alberto Bressan,** Lecture Notes on Functional Analysis: With Applications to Linear Partial Differential Equations 9 787040 569773

107 **Wilhelm Schlag,** A Course in Complex Analysis and Riemann Surfaces 9 787040 569797

108 **Emmanuel Kowalski,** An Introduction to the Representation Theory of Groups 9 787040 569780

109 **Steven G. Krantz,** A Guide to Complex Variables 9 787040 570229

110 **Rajendra Bhatia,** Fourier Series 9 787040 570236

111 **Fernando Q. Gouvêa,** A Guide to Groups, Rings, and Fields 9 787040 569995

112 **Steven G. Krantz,** A Guide to Functional Analysis 9787040570243

113 **Alberto Torchinsky,** Problems in Real and Functional Analysis 9787040570250

114 **Fritz Colonius,Wolfgang Kliemann,** Dynamical Systems and Linear Algebra 9787040570212

115 **Jerome K. Percus, Stephen Childress,** Mathematical Models in Developmental Biology 9787040570342

116 **Steven G. Krantz,** A Panorama of Harmonic Analysis 9787040570274

117 **Ralph P. Boas, Revised by Harold P. Boas,** Invitation to Complex Analysis, Second Edition 9787040570281

118 **William H. Meeks III, Joaquín Pérez,** A Survey on Classical Minimal Surface Theory 9787040570298

119 **Peter Duren,** Invitation to Classical Analysis 9787040570304

120 **Helene Shapiro,** Linear Algebra and Matrices: Topics for a Second Course 9787040570311

121 **Ralph P. Boas, Revised and updated by Harold P. Boas,** A Primer of Real Functions, Fourth Edition 9787040570328

122 **John P. D'Angelo,** Inequalities from Complex Analysis 9787040570335

123 **Terence Tao,** Expansion in Finite Simple Groups of Lie Type 9787040592979

124 **Goro Shimura,** Arithmetic and Analytic Theories of Quadratic Forms and Clifford Groups 9787040592986

125 **Victor M. Buchstaber, Taras E. Panov,** Toric Topology 9787040592993

126 **Benson Farb, Richard Hain, Eduard Looijenga,** Moduli Spaces of Riemann Surfaces 9787040593099

127 **Charles Livingston,** Knot Theory 9787040593075

128 **Mladen Bestvina, Michah Sageev, Karen Vogtmann,** Geometric Group Theory 9787040593105

129 **Edward Packel,** The Mathematics of Games and Gambling, Second Edition 9787040593082

130 **Kiran S. Kedlaya, Bjorn Poonen, Ravi Vakil,** The William Lowell Putnam Mathematical Competition 1985–2000: Problems, Solutions, and Commentary 9787040593129

131 **Ida Kantor, Jiří Matoušek, Robert Šámal,** Mathematics++: Selected Topics Beyond the Basic Courses 9787040593136

132 **John Roe,** Winding Around: The Winding Number in Topology, Geometry, and Analysis 9787040593143

133 **Wolfgang Kühnel, Translated by Bruce Hunt,** Differential Geometry: Curves–Surfaces–Manifolds, Third Edition 9787040593150

134 **Daniel W. Stroock,** Mathematics of Probability 9787040593020

135 **Jennifer Schultens,** Introduction to 3-Manifolds 9787040593037

136 **Gábor Székelyhidi,** An Introduction to Extremal Kähler Metrics 9787040593044

137 **Diane Maclagan, Bernd Sturmfels,** Introduction to Tropical Geometry 9787040593051

138 **Barry Simon,** Real Analysis : A Comprehensive Course in Analysis, Part 1 9787040593068

139 **Barry Simon,** Basic Complex Analysis: A Comprehensive Course in Analysis, Part 2A 9787040593006

140 **Barry Simon,** Advanced Complex Analysis: A Comprehensive Course in Analysis, Part 2B

141 **Barry Simon,** Harmonic Analysis: A Comprehensive Course in Analysis, Part 3

142 **Barry Simon,** Operator Theory: A Comprehensive Course in Analysis, Part 4

143 **Sergei Ovchinnikov,** Number Systems: An Introduction to Algebra and Analysis

144 **Ian M. Musson,** Lie Superalgebras and Enveloping Algebras

145 **John B. Conway,** A Course in Abstract Analysis

146 **V. I. Arnold,** Lectures and Problems: A Gift to Young Mathematicians

147 **Pascal Cherrier, Albert Milani,** Linear and Quasi-linear Evolution Equations in Hilbert Spaces

148 **John M. Lee,** Axiomatic Geometry

149 **Shiri Artstein-Avidan, Apostolos Giannopoulos, Vitali D. Milman,** Asymptotic Geometric Analysis, Part I

150 **Gregory Berkolaiko, Peter Kuchment,** Introduction to Quantum Graphs

151 **Libor Šnobl, Pavel Winternitz,** Classification and Identification of Lie Algebras

152 **Maciej Zworski,** Semiclassical Analysis

153 **Thomas Garrity, Richard Belshoff, Lynette Boos, Ryan Brown, Carl Lienert, David Murphy, Junalyn Navarra-Madsen, Pedro Poitevin, Shawn Robinson, Brian Snyder, Caryn Werner,** Algebraic Geometry: A Problem Solving Approach

154 **Luis Barreira, Yakov Pesin,** Introduction to Smooth Ergodic Theory

155 **László Lovász,** Large Networks and Graph Limits

156 **Jean-Marie De Koninck, Florian Luca,** Analytic Number Theory: Exploring the Anatomy of Integers

157 **María Cristina Pereyra, Lesley A. Ward,** Harmonic Analysis: From Fourier to Wavelets

158 **Gail S. Nelson,** A User-Friendly Introduction to Lebesgue Measure and Integration

159 **Mark Kot,** A First Course in the Calculus of Variations

160 **Michael Aizenman, Simone Warzel,** Random Operators: Disorder Effects on Quantum Spectra and Dynamics

161 **Charles A. Weibel,** The K-book: An Introduction to Algebraic K-theory

162 **Aaron N. Siegel,** Combinatorial Game Theory